BIRDS AND BIBLES
IN HISTORY

By:

TIAN HATTINGH

Eurasian Eagle-owl *(Bubo bubo)*
(Photo: Gareth Poulton)

The
London
Press

First published by The London Press, UK 2011

Disclaimers & Credits: Save for the images of John Denver, my parents, grandpa, and myself, all the images in this publication are reproductions of files from www.flickr.com, and the Wikimedia Commons. I would like to use this opportunity to thank all the copyright holders for their generosity in sharing their work from www.flickr.com (35 members, contributing 85 photographs) and in the Wikimedia Commons (5 members, each contributing a photograph). A number of these photographs were used to produce the icons used in the headers, divisions, and chapter endings.

This work may contain copyrighted material the use of which has not always been specifically authorised by the copyright owner. Such material is made available for educational purposes, to advance understanding of human rights, democracy, scientific, moral, ethical, and social justice issues, etc. It is believed that this constitutes a 'fair use' of any such copyrighted material as provided for in Title 17 U.S.C. section 107 of the US Copyright Law.

Front Cover: White-bellied Sunbird *(Cinnyris talatala)*. This image is a reproduction of the photograph taken by Rus Koorts from Pretoria, South Africa, and used with permission. Originally placed on www.flickr.com, the original photograph can be viewed at: http://www.flickr.com/photos/ruslou/4821493124/in/set-72157623311955116. The other images are from the Title Page, Par. 3.2, and Par. 23.1. Please refer to the attributions there.

Back cover: These images are from Par. 3.1, PART 4 front page, Par 40.8, Par. 16.1, Par. 6.1.1. Please refer to the attributions there.

Title Page: This image is a reproduction of the photograph taken by Garreth Huw Photography and used with permission. Originally placed on www.flickr.com, the original photograph can be viewed at: http://www.flickr.com/photos/polts23/424407402/.

Inner back page: THE HOLY BIBLE, NEW INTERNATIONAL VERSION®, NIV® Copyright © 1973, 1978, 1984, 2011 by Biblica, Inc.™ Used by permission. All rights reserved worldwide.

I would like to hear from you. Please visit my BBH website at:
http://birds-and-bibles-in-history.com/ for more news on the subjects covered in the book, and an opportunity to state your views in the forums.
(Hierdie webwerf beskik ook oor 'n Afrikaanse afdeling.)

Perfectbound: 978-1-907313-70-7
Ebook: 978-1-905006-46-5

A C.I.P reference is available from the British Library

There are three things which are too wonderful for me,
Yes, four which I do not understand:
The way of an eagle in the air,
The way of a serpent on a rock,
The way of a ship in the midst of the sea,
And the way of a man with a virgin.

Proverbs 30:18, 19 (NKJV)

CONTENTS

Preface xiii

Acknowledgements xvi

Abbreviations..xix
    a) Relevant Bible Books xix
    b) General..xix

Introduction xxi
    a) Objectives of this Book..xxi
    b) Target Population xxii
    c) How to use this publication xxiii
        i) Seeking an answer to the hypothesis xxiii
        ii) From an Old Testament text to a bird..................xxiii
        iii) From a bird to a verse xxv
        iv) Unidentified birds..xxv
        v) The so-called "unclean" birds xxv
    d) I Believe in… xxv
    e) John Denver xxviii

Dedication xxxii

PART I ORNITHOLOGICAL HISTORY

Chapter 1 ORNITHOLOGICAL HISTORY 2
  1.1 In the Beginning.. 2
  1.2 The Old Testament Authors 5
      1.2.1 Moses 5
      1.2.2 Job 5
      1.2.3 Isaiah 6
      1.2.4 Jeremiah... 6

1.3 Aristotle 7
1.4 A Chinese Encyclopedia 9
1.5 Pliny the Elder 9
1.6 The Middle Ages 10
    1.6.1 The Early Middle Ages (476–1000) 11
    1.6.2 The High Middle Ages (1000–1300) 12
    1.6.3 The Late Middle Ages (1300–1453) 13

Chapter 2   THE RENAISSANCE 15
2.1 Contemporaries 15
    2.1.1 The Fourteenth Century 15
    2.1.2 The Fifteenth Century 16
    2.1.3 The Sixteenth Century 19
    2.1.4 The Seventeenth Century 21
    2.1.5 The Eighteenth Century 21
2.2 John Ray 24
2.3 Carl von Linne 26
2.4 Gilbert White 27
2.5 Daines Barrington 28
2.6 Lazzaro Spallanzani 29
2.7 Charles Waterton 30

Chapter 3   THE MODERN ERA 31
3.1 John James Audubon 31
3.2 Charles Darwin 34
    3.2.1 A Personal Note 36
3.3 Edward Lear 38
3.4 The Behaviourists 38
3.5 Ethology 40
3.6 Animal societies 42

Chapter 4   MODERN DEVELOPMENTS 45
4.1 Introduction 45
4.2 Migration 47
    4.2.1 The Great Migration 49
    4.2.2 The Lepidoptera Migrations 51
    4.2.3 The Salmon Runs 52
4.3 Bird Ringing / Banding 52

4.4 Anti-strike Units 54

4.5 Extinction 55

4.6 Nomenclature 59

    4.6.1 The Binominal System 59

    4.6.2 The Living World 60

    4.6.3 The Class: Aves 60

4.7 Taxonomy 61

4.8 Palestinian Zoogeography 62

PART 2 BIBLE TRADITIONS

Chapter 5 BIBLES IN HISTORY 66

5.1 Chronology 66

5.2 The Holy Bible 67

5.3 The Jewish Bible 68

    5.3.1 The Torah 69

    5.3.2 The Nevi'im 70

    5.3.3 The Ketuvim 70

5.4 The Ancient Versions 71

    5.4.1 The Targums 72

    5.4.2 The Septuagint 72

    5.4.3 The Vulgate 73

    5.4.4 The Masoretic Texts 74

    5.4.5 The Dead Sea Scrolls 74

5.5 The Christian Bibles 75

    5.5.1 The Old Testament 76

    5.5.2 The New Testament 77

    5.5.3 The History of Bible Translations 79

    5.5.3.1 Early Translations in Late Antiquity 79

    5.5.3.2 The Middle Ages 79

    5.5.3.3 Reformation and Early Modern Period 80

Chapter 6 TEXTUAL CRITICISM 83

6.1 The Bible? 83

    6.1.1 The Time Factor 84

    6.1.2 The Manuscripts 84

6.2 In Conclusion 85

6.3 An illustration 86

6.4 In a Nutshell 87

PART 3   BIBLICAL BIRDS

Chapter 7   ORDER: STRUTHIONIFORMES 90

7.1 Flightless Birds 90

Chapter 8   FAMILY: STRUTHIONIDAE 92

8.1 The Ostrich 92

8.2 Utilization by Man 95

8.3 Biblical References 96

Chapter 9   ORDER: FALCONIFORMES 97

9.1 Diurnal Birds of Prey 97

9.2 Non-Biblical Families 98

9.3 Biblical Families 99

Chapter 10   FAMILY: ACCIPITRIDAE 102

10.1 Eagles 102

10.1.1 Biblical References 105

10.2 Vultures 106

10.2.1 Biblical References 108

10.3 Kites 109

10.3.1 Biblical References 110

Chapter 11   FAMILY: FALCONIDAE 112

11.1 Falcons... 112

11.2 Falconry 114

11.3 Biblical References 115

Chapter 12   FAMILY: PANDIONIDAE 116

12.1 The Osprey 116

12.2 Ospreys in Britain 118

12.3 Biblical References 119

Chapter 13   ORDER: GALLIFORMES 120
13.1 Fowl-like Birds 120
13.2 Non-Biblical Families 122

Chapter 14   SUB ORDER: PHASIANOIDEA
FAMILY: PHASIANIDAE 123
14.1 Introduction 123
14.2 Partridges 124
   14.2.1 Biblical References 125
14.3 Quail 125
   14.3.1 Biblical References 126
14.4 Red Junglefowl 127
   14.4.1 Biblical References 128
14.5 Indian Peafowl 129
   14.5.1 Biblical References 130

Chapter 15   ORDER: STRIGIFORMES 132
15.1 Nocturnal Raptors 132
15.2 Classification 135

Chapter 16   FAMILY: TYTONIDAE 136
16.1 Common Barn-owl 136
16.2 Non-Biblical References 140
16.3 Biblical References 141

Chapter 17   FAMILY: STRIGIDAE 142
17.1 Typical Owls 142
17.2 Eurasian Eagle-owl 144
   17.2.1 Biblical References 145
17.3 Eurasian Scops-owl 145
   17.3.1 Biblical References 146
17.4 Pallid Scops-owl 146
   17.4.1 Biblical References 147
17.5 Little Owl 147
   17.5.1 Biblical References 148
17.6 Northern Long-eared Owl 148
   17.6.1 Biblical References 149
17.7 Short-eared Owl 149

17.7.1 Biblical References 150
17.8 Non-biblical Species 150

Chapter 18   ORDER: COLUMBIFORMES 152
18.1 The Order 152

Chapter 19   FAMILY: COLUMBIDAE 155
19.1 Pigeons and Doves 155
19.2 Non-Biblical References 157
19.3 Biblical References 157

Chapter 20   ORDER: PASSERIFORMES 160
20.1 The Perching Birds / Songbirds 160

**Chapter 21   SUBORDER: PASSERI (OSCINES)
FAMILY: HIRUNDINIDAE** 163
21.1 Swallows 163
21.2 Biblical References 165

Chapter 22   FAMILY: CORVIDAE 166
22.1 Ravens and Crows 166
22.2 Raven 167
22.3 Biblical References 168

Chapter 23   FAMILY: PLOCEIDAE 169
23.1 Sparrows 169
23.2 Biblical References 170

Chapter 24   ORDER: APODIFORMES 172
24.1 Swifts, Tree Swifts, and Hummingbirds 172

Chapter 25   TRIBE: APODINI 173
25.1 Typical Swifts 173
25.2 Biblical References 176

Chapter 26   GENERAL AND UNRESOLVED 177
26.1 Bats 177
26.1.1 Biblical References 178

26.2 Unidentified Birds 179
  26.2.1 Single Old Testament Occurrences 179
  26.2.2 Double Old Testament Occurrences 180
26.3 Unresolved Texts 181
  26.3.1 Isaiah 34:11 181
  26.3.2 Job 40:24–29 182
  26.3.3 Jeremiah 8:7 182
  26.3.4 Ezekiel 13:20 183

PART 4   CONCLUSIONS

Chapter 27   CONCLUSIONS 186
27.1 Concluding Statement 186
27.2 Genesis 6–8 187
27.3 Job 39–41 188
27.4 The Prophets 189
27.5 The Future 189

Chapter 28   POSTSCRIPT 190
28.1 Postscript 190

Chapter 29   BIBLIOGRAPHY 192
29.1 Bibles 192
29.2 Dictionaries 193
29.3 Commentaries 193
29.4 Birds 194
29.5 Biblical Hebrew 195
29.6 General 196

PART 5   APPENDIXES

Chapter 30   CLASSIFICATION: THE LIVING WORLD 200

Chapter 31   CLASSIFICATION: BUBO AFRICANUS 201

Chapter 32   REFERENCES TO BIRDS PER BIBLE BOOK 202

32.1 Old Testament books 203
32.2 New Testament books 207

Chapter 33   REFERENCES TO AVIAN TERMS 208
33.1 Old Testament 208
33.2 New Testament 209

Chapter 34   REFERENCES TO BIRDS IN GENERAL 210
34.1 Old Testament 210

Chapter 35   BIBLE SPECIES ALPHABETICAL 212
35.1 Old Testament 212
35.2 New Testament 213

Chapter 36   REFERENCES AS PER BIBLE SPECIES 214
36.1 Old Testament 214
        36.2 New Testament 217

Chapter 37   INDEX TO SPECIES AND FAMILIES 219
37.1 Old Testament species and families 219
37.2 Other Species and Families 222

Chapter 38   UNIDENTIFIED BIRDS 230
38.1 Single Old Testament Occurrences 230
38.2 Double Old Testament occurrences 231

Chapter 39   "UNCLEAN" BIRDS 232
39.1 Leviticus 232
39.2 Deuteronomy 233

Chapter 40   BIRDING IN ISRAEL 234
40.1 Introduction 234
40.2 Chutzpah 236
40.3 Parks and Reserves 237
40.4 Infrastructure 237
40.5 Military Zones 238
40.6 Northern Region 239
40.7 Central Region 240

40.8 Southern Region 241

Psalm 8 243

Biography 244

Brown Pelican *(Pelecanus occidentalis)*
(Photo: Ingrid Taylar) [1]

---

[1] This image is a reproduction of the photograph taken by Ingrid Taylar, Seattle, WA, USA and used with permission. Originally placed on www.flickr.com, the original photograph can be viewed at: http://www.flickr.com/photos/taylar/4101640679/

PREFACE

My parents were nature and outdoor lovers. Especially my father. So from an early age I was made aware of the splendors of Mother Nature. Our holidays were always spent caravanning in the great outdoors. After completing my schooling I had to spend a year in the military and was fortunate enough to be posted to Walvis Bay, a coastal town on the edge of the Namib Desert in the present day Namibia. After that, I spent two years as a Pupil Forester at the "Saasveld Forestry College" as it was then called. (The Port Elizabeth Technicon expanded to the Southern Cape in 1985 when it took over the

Charel and Margaret Hattingh
(Photo: Tian Hattingh) [2]

college from the Department of Forestry. In 2005 it became part of the Nelson Mandela Metropolitan University). There I met Dave Pepler, and because of his enthusiasm, I became interested in the birds and in the different aspects of ornithology. He later became a respected scientist and academic with an M. Phil. in Zoology from Cambridge. He is currently presenting the environmental program Groen (Green) on the TV channel kykNET and is now one of South Africa's best-known environmentalists.

In July of 1990 AD, I visited the Holy Land as a Christian pilgrim. The worth of this visit was twofold: Firstly, the text of the Bibles which I was reading and studying since childhood was to a large extent replaced by real images and experiences. Secondly, it stirred in me the longing to return and explore the birdlife of the region. Later I involved myself in studying undergraduate Biblical Hebrew at the University of South Africa. One of my teachers was Prof. J.C. Lübbe, Head of the Department of Semitics. He not only taught his students the language, but opened our eyes to a very special way of dealing with the text of the Old Testament.

The fulfillment of my dream to re-visit the Middle East became a reality much sooner than I had anticipated, as I was given the opportunity to visit Israel again in

---

[2]   This image is a photograph from my own collection.

March/April 2000. This time around, Lesley Ryan, my good friend and fellow birder at the then Rustenburg Bird Club, (now BirdLife Rustenburg) in the North-West Province of South-Africa, accompanied me. We had the privilege of traveling three thousand kilometers at our own will, visiting virtually all the worthwhile locations from a birding point of view, and experiencing the country and its people first hand. As I observed the country and its birdlife, I became more and more convinced that the birds in my Bibles were very special, and that my knowledge on the subject was extremely limited. This prompted me to search my Bibles for references to birds, and the exercise in turn resulted in me realizing that the subject was a fascinating, but also an extremely complex one. As I discovered more and more interesting, and even mind-boggling, facts, dilemmas, and unsolved mysteries, the urge to share my findings with others began to emerge. The idea to write this book started taking shape, and I returned to Israel from November 2000 to February 2001. Using Jerusalem as my base, I gained further information and insights into the subject matter.

Following is a summary of the references to birds and avian terms that are dealt with in this book:

	OT	NT	Total	Chapters
Identified species	174	0	174	32 35 36
Refer to birds in general	108	47	155	32 34
Avian terms	61	19	80	33
Total	**343**	**66**	**409**	

These references can be found in the sixty-six books of the modern Christian Bibles, including thirty-six species in seven Orders. Details are given in the different Indexes. (See par. 32–39). The events described in the Masoretic texts of the Old Testament took place in the Middle East. It was recorded by the authors and/or compilers of the books, and later copied and edited by scribes in that and neighboring areas. For the purposes of this book, it's the birds found in this area that interest us.

As for the origins and development of Ornithology, I realize that my research has been largely Euro-centric. The lack of documentation in English from the other continents make it virtually impossible for me to establish if any significant developments indeed took place and if so, when and where. After living and working in the mainland of China for the past nine years (2002–2011), I felt the urge to include par. 1.4 to indicate that, while Ornithology had been developing in areas known to me, there have been a significant number of developments in places that I have not even thought of.

White Stork (*Ciconia ciconia*)
(Photo: Ted van den Bergh) [3]

---

[3] This image is a reproduction of the photograph taken by Ted van den Bergh and used with permission. Originally placed on www.flickr.com, under the username "webted", the original photograph can be viewed at: http://www.flickr.com/photos/webted/445159012/

ACKNOWLEDGEMENTS

Because of the unique topic of this study, I have to a large extent been alone in the writing phase of the book, but many people have contributed in one way or another in me reaching this point. Behind each one of the names below, an interesting story of unselfish friendship or professional assistance can be told. I would hereby like to salute each and every one of them.

ISRAEL:
Kibbutz Kfar Ruppin: David Glasner, Elana, Czech, Mordegai, and Amotz.
Jerusalem: John and Amanda van der Walt; Pieter and Edelene Marais; Pieter and Babs Voster.
Kibbutz Lotan: David Dov and his colleagues.
Eilat: Dr. Reuben Josef.

SOUTH AFRICA:
Rustenburg:
Lesley Ryan, a dear friend, for the unlimited use of her extensive birding library.
Marjo Kruger, Chairman of BirdLife Rustenburg, for being a friend indeed.
A special word of thanks to all the members of BirdLife Rustenburg who I have had the privilege of knowing as friends and with whom I have spent many wonderful times birding throughout South and Southern-Africa. "Moshe will always remember you."

Krugersdorp:
A special word of thanks has to be addressed to my parents, Charel and Margaret Hattingh. They were obedient to their vows at my baptism, and raised me, and their other children, in the ways of our Lord, Jesus Christ. Through their love for the outdoors and nature, I had the privilege of visiting many of the exceptional places throughout South- and Southern-Africa from a young age onwards. They were instrumental in making one of my longer visits to the Middle East financially possible. For this I would always be grateful.

Johannesburg:
Dr. Aldo Berruti: Member, Birdlife International Global Council, (1998–present), Member, International Ornithological Com, (1998–present), Chief Executive, BirdLife SA, (1996–present).
Jeff Lockwood, Delta Park Environmental Center.
Jannie du Plessis: Director, Pilgrimage Tours.
My brother, Hein-Phillip, and his lovely wife Tracey.

Pretoria:
Prof. J.C. Lübbe: Head, Department of Semitics, The University of South Africa.

CHINA:
Shenzhen: Ajax Zhang, without whose help this publication would never have materialized.

UNITED KINGDOM:
London: Simon Kelly and his team at The London Press, for their patience and encouragement in making this publication what it is. Without their assistance it would not have been at all possible.

WIKIPEDIA (www.wikipedia.com)
Wikipedia is a free, web-based, collaborative, multilingual encyclopedia project supported by the non-profit Wikimedia Foundation. Its eighteen million articles (over 3.6 million in English) have been written collaboratively by an army of volunteers from all around the world. Almost all of its articles can be edited by anyone with access to the site. Wikipedia was launched in 2001 by Jimmy Wales and Larry Sanger and has become the largest and arguably the most popular general reference work on the Internet. It ranks around seventh among all websites, and has in the order of 365 million readers. The bulk of the text of this book was completed in 2000 before Wikipedia was launched. It was revised ten years later, and then Wikipedia proved to be a mind-boggling source of information.

HANDBOOK OF THE BIRDS OF THE WORLD
In the Wikipedia article "Handbook of the Birds of the World" it is said: "*The Handbook of the Birds of the World* (HBW) is a multi-volume series produced by the Spanish pub-

lishing house Lynx Edicions. It is the first handbook to cover every known living species of bird. The series is edited by Josep del Hoyo, Andrew Elliott, Jordi Sargatal, and David A. Christie."[4]

New volumes appear at annual intervals, and the series is expected to be complete with Volume 16 by November 2011. When Volume 16 is published, for the first time an animal class will have all the species illustrated and treated in detail in a single work. This has not been done before for any other group in the animal kingdom. During the writing of this publication, I have consulted Volumes 1–5 extensively. The wealth of information is nothing less than a birding library, and I am grateful for the editors for their vision, courage, and determinism in completing this mammoth task. Who knows, this series might in future become known as "The Bible of Birds."

White-Faced Ducks (*Dendrocygna viduata*)
(*Photo: Rus Koorts*) [5]

---

[4]   Wikipedia contributors. "Handbook of the Birds of the World." Wikipedia, The Free Encyclopedia. Wikipedia, The Free Encyclopedia, 21 Mar. 2011. Web. 7 Aug. 2011.

[5]   This image is a reproduction of the photograph taken by Rus Koorts, Pretoria, South Africa and used with permission. Originally placed on www.flickr.com, the original photograph can be viewed at: http://www.flickr.com/photos/ruslou/6055084229/.

ABBREVIATIONS

a) Relevant Bible Books

Old Testament

Gen	Genesis	Ps	Psalms	Hos	Hosea
Ex	Exodus	Prv	Proverbs	Am	Amos
Lev	Leviticus	Ecl	Ecclesiastes	Obd	Obadiah
Num	Numbers	So	Song of Songs	Mi	Micah
Deut	Deuteronomy	Is	Isaiah	Nah	Nahum
1 Sam	1 Samuel	Jer	Jeremiah	Hab	Habakkuk
1 Kng	1 Kings	Lam	Lamentations	Zeph	Zephaniah
2 Chr	2 Chronicles	Ezk	Ezekiel	Zech	Zechariah
Job	Job	Dan	Daniel		

New Testament

Mat	Matthew	Jhn	John	I Cor	I Corinthians
Mrk	Mark	Act	Acts	Rev	Revelation
Luk	Luke	Rom	Romans		

b) General

AD	*Anno Domini*. Latin for "In the Year of (Our) Lord." After Christ.
A33	The Bible in Afrikaans (1933 edition)
A83	The Bible in Afrikaans (1983 edition)
BC	Before Christ
BHS	Biblia Hebraica Stuttgartensia
JB	Jerusalem Bible
LB	Living Bible
MSG	The Message
NASB	New American Standard Bible

NIV   The Holy Bible (New International Version)
NKJV   New King James Version
RSV   Revised Standard Version
TEV   Good News Bible (Today's English Version)

Red Kite *(Milvus milvus)*
(Photo: Gareth Paulton) [6]

---

[6] This image is a reproduction of the photograph taken by Gareth Paulton and used with permission. Originally placed on www.flickr.com, the original photograph can be viewed at: http://www.flickr.com/photos/polts23/520228167/in/set-72157600282480789.

INTRODUCTION

a) Objectives of this Book

The objectives of this publication could be described as follows:

- Firstly. To present the reader with as much relevant, noteworthy, and otherwise interesting information as possible, regarding the following three topics:

  - The development of Ornithology from ancient to modern times, exploring its origins, noting the landmarks reached along the way, and briefly evaluating the work, achievements, and discoveries of the main characters involved.

  - The birds mentioned in the Christian Bibles, in other words the Masoretic text of the thirty-nine books of the Christian Old Testament Canon, and the Greek texts of the twenty-seven books of the Christian New Testament. This would include general information on the Orders and Families to which these birds belong, and where it is possible to identify a specific species beyond any reasonable doubt, present information on those species. The latter would include concise notes on the birds' morphological features, its distribution on a worldwide scale, its habitat preferences, general habits, breeding, feeding, and, where applicable, its relationship to man.

  - The information gained from the above activities will be utilized to achieve the next objective. Because somebody, somewhere had to be the first to practice ornithology, my quest was to determine, beyond any reasonable doubt, which is or are to be considered the "original Ornithologist(s)", and where the "true origins of Ornithology" is to be found. For this I have formulated a hypothesis, and in the end reached a conclusion on its validity. The hypothesis reads as follows:

> "The Hebrew Canon,
> as expounded in the Mosaic Law, the Prophets, and the Writings,
> contains the origins of Ornithology,
> and the Biblical authors concerned, and not Aristotle,
> were the original Ornithologists."

- Secondly. During 1973, I was studying at the University of Stellenbosch, South-Africa, where I attempted to obtain a Bachelors degree in Forestry. I was unsuccessful in getting the degree but gained something that later in my life proved to be much more valuable. During a lecture, our Chemistry professor's thoughts apparently strayed for a moment, and as he looked at our young, inexperienced faces, he almost pleadingly exclaimed: "Ladies and gentleman, you should read the Good Book, but please, read some others as well." Even if this book were able to prompt only one person to start, or at least continue, doing just that, I would then be able to declare that the book has made a difference, and so fulfilled a worthy cause.

- Lastly. I envisage that the book would serve as a source of reference on the data pertaining to all bird species mentioned, and avian terminology employed, in the Masoretic Hebrew texts as found in the Christian Old Testament Canon. By presenting complete indexes in different formats, the reader/scholar will have the ability to locate information easily.

b) Target Population

This book is aimed at those individuals, who have, to some degree or another, become interested in the world of birds, and who have, when reading their Bibles, noticed the birds in the texts, and wanted to know more about them. As with many other activities in life, the extent to which an individual amateur birder is able to practice his or her hobby is often determined by the financial abilities of the participant. Birding in far-off and exotic places is simply not within everyone's reach. So, in order to give the reader a global perspective of the relevant families, I have gone to great lengths in describing the relative positions of the Families in the Orders, as well as a general overview of the Groups or Tribes in the families. It is my sincere wish that this book would to some extent satisfy the desire of those who would love to experience the Middle East and its birds for themselves, but are not able to do so. In order to keep the book reader friendly, I have at all times been weary not to let it become a purely scientific publication, although I have endeavored to present it as professionally as possible, with the means at my disposal. In order to keep the format of the book manageable, only seven of the fifteen Orders in which Bible Birds are found have been dealt with in detail. The latter was done with the purpose of giving the reader a wider perspective on where the Bible Birds fit into the global context, and to enhance the books' worth as a reference work, making it useful to biblical as well as fledging (pun intended) ornithological students.

c) How to use this publication

i) Seeking an answer to the hypothesis:

One of the main objectives of the book as stated in par. a) is "to determine, beyond any reasonable doubt, which is or are to be considered the 'original Ornithologist(s)', and where the 'true origins of Ornithology' are to be found." To start off, I would suggest that one should read the book from beginning to end. This will lead the reader through all the relevant information and eventually to the "Conclusion" in Chapter 27. It will show whether the hypothesis as stated is indeed true or not.

Common Barn-owl *(Tyto alba)*
(Photo: Paul Jeapes) [7]

ii) From an Old Testament text to a bird.

One finds a bird or an avian term mentioned in the Old Testament of any Bible and wonders what exactly is meant. The following steps will assist the reader in finding information quick and easy:

---

[7]   This image is a reproduction of the photograph taken by Paul Jeapes and used with permission. Originally placed on www.flickr.com, the original photograph can be viewed at: http://www.flickr.com/photos/pauljeapesphotography/5787854664/.

STEP 1: Go to Chapter 32 to find the location of all references in a Bible Book, or Chapter 33 to find the location of avian terms, or Chapter 34 to search for the location of references to birds in general.

NOTES:

1. If the name or term searched for is not included in these lists, it means that the original Hebrew text does not refer to a bird or birds in general, or an avian term directly. In other words, the translators of the Old Testament that the reader is using have decided to use the name of a bird or an avian term according to their own interpretation of the particular text.

2. Always keep in mind that the numbering of verses differs in different translations. Readers should check one or even two verse numbers before and after the verse number used in the Bible they are currently using.

3. The names and terms used in Chapters 32, 33, and 34 are the **suggested translations** that the author has reached after considering all the relevant information, options and opinions. These are by no means intended to be the "final" or "correct" translation.

Eurasian Eagle Owl *(Bubo bubo)*
(Photo: Rus Koorts) [8]

STEP 2: Go to Chapter 35 where thirty-six biblical species that can be identified positively are listed alphabetically according to the English names, and then to Chapter 36 to find all the other references to the specific Bible species.

STEP 3: Go to Chapter 37 to find the paragraphs in this publication that deals with a particular name and Chapter 38 if the bird or term cannot be identified beyond any reasonable doubt.

---

[8] This image is a reproduction of the photograph taken by Rus Koorts from Pretoria, in South Africa and used with permission. Originally placed on www.flickr.com, the original photograph can be viewed at: http://www.flickr.com/photos/ruslou/2292597657/sizes/o/in/set-72157603969374345/.

iii) From a bird to a verse:

STEP 1: Go to Chapter 35 where Bible species that can be identified positively are listed alphabetically according to the English names, and then to Chapter 36 to find all the other references to the specific Bible species.

STEP 2: Go to Chapter 37 to find the paragraphs in this publication that deals with this bird or avian term, and Chapter 38 if the bird cannot be identified beyond any reasonable doubt.

iv) Unidentified birds.

Chapter 38 deals with the tough cases where we cannot find much to assist us in making a calculated guess as to which species is meant. I believe that the work being done at the moment by researchers will shed new light on many of these texts. For example: at the University of South-Africa under the leadership of Professor J.C. Lübbe, a team is compiling *A semantic domains dictionary of the Old Testament*, which will give us many new insights.

v) The so-called "unclean" birds.

Chapter 39 deals exclusively with the so-called "unclean" birds mentioned in Leviticus 11 (Refer to par. 39.1), and Deuteronomy 14 (Refer to par. 39.2).

d) I Believe in …

I believe in tolerance. In 1903 Hellen Keller in Part II of her essay entitled *Optimism* said that the highest result of education is tolerance. I can only hope that this publication will be educational and result in more tolerance. Throughout this publication I will make some statements that are, to say the least, highly controversial. I fully realize that there is a wide continuum of viewpoints on each and every one of these issues. I do not expect anyone to accept my point of view, but only ask that they would keep an open mind, and at least take note of my and other opinions. I surely do.

> "I believe in you and me. I'm like Albert Schweitzer and Bertrand Russell and Albert Einstein in that I have a respect for life – in any form. I believe in nature, in the birds, the sea, the sky, in everything I can see or that there is real evidence for. If these things are what you mean by God, then I believe in God."
>
> **Frank Sinatra (1915–1998)**
> *(American actor and singer)*

Throughout the history of man, more wars and conflict have been caused by religious intolerance than any other single reason. The Middle East has been the cradle of three of the major religions of the world namely: Christianity, Islam and Judaism. In Genesis Chapter 12, the calling of the patriarch Abraham and his subsequent move to the area is the start of the Old Testament's narration that would cover many centuries. It tells the turbulent history of the "People of God" in this part of the world. Even today there are many outstanding religious, political and other issues, resulting in hatred, conflict, violence and bloodshed. Therefore, when visiting countries in the region, or writing a book such as this, one has to be extremely careful not to offend one or another cultural group and/or religious grouping and/or political view. I could only hope that this book would be acceptable to members of a wide range of people from all communities and religious persuasions as I accept them as they are, and they hopefully accept me as I am. As a Christian I live according to my interpretation of what I believe God prescribes in the so-called "Christian Bibles." Jesus summarized these prescriptions as love towards God and fellow man. To me, this would invariably include religious and political freedom and tolerance. Every human being should be free to worship who s/he wants, where, when and how s/he wants, providing that no other person is affected negatively when doing so. I am convinced that freedom should always be accompanied by responsibility and accountability. If not, freedom becomes a type of slavery in disguise, as the lives of extremists clearly demonstrate.

In other words, I believe that no one needs to apologize for what s/he believes in. I certainly don't. As long as one respects those who believe differently, or those who do not believe at all. For example: I have been living and working as an English teacher in the People's Republic of China for the past nine years (2002–2011). To me it is completely understandable that when people hear what I believe in, they find it strange to say the least, and sometimes, even comical. I have found countless people who were completely oblivious of the fact that there are actually books called "Bibles." The Old Testament narrations concerning the "People of God" are often unbelievable to them in the true sense of the word. The New Testament describing a virgin giving birth to a baby boy is a case in point. A common, uneducated man, who wrote nothing other than a few words in the sand, calling himself the "Son of God" is definitely frowned upon. Things like miracles, angels, demons, and corpses springing back to life. Concepts like sin, heaven, hell, redemption, salvation, faith, and what not. These are largely unknown phenomena, often absolutely astonishing, and therefore completely alien to their mind-set and world-views. It is therefore imperative that as a believer one accepts this *status quo* as a simple fact of life, and only then can one live, work, and play harmoniously together and even become friends.

I believe in democracy in the widest, most basic sense of the word. I believe that it means different things to different people in different places, as determined by differences in cultures, historic backgrounds, socio-economic circumstances, and a host of other factors. Although I recognize its shortcomings, I believe that the capitalist system is the most advantageous to the largest majority of people at the present moment in the history of our planet. Poverty, starvation, disease, ignorance etc. are the results of the greed found to some extent in all of us, and to a large extent, in many of us. The wealthy often justify their position by arguing that what they had achieved had been as a result of hard work and/or taking risks. I would respect that if the following provisos were adhered to in the process of accumulating their wealth: a) no exploitation of others had taken place, b) no crime or corruption had been committed, and c) if they would admit that the extent of their talents and energy were not of their own doing. The poor often lament their position by adopting a deterministic world-view. By now (2011) it is clear that the widening gap between the wealthy and the poor (individuals and nations) constitutes a significant part of the challenges facing us humans in the near future. However, having said that, I also believe that we as believers should remind ourselves, that not a single one of the founders of a major religion extant today (Jewish, Muslim, Hindu, Christian, Taoist, Buddhist, etc.) were capitalists.

> "As the eagle was killed by the arrow winged with his own feather, so the hand of the world is wounded by its own skill."
>
> **Helen Keller (1880–1968)**
> *(American Author and Educator who was blind and deaf)*

I believe that greed is fast destroying our planet (See par. 4.5). The extent to which, for example, over population, global warming, and the unsustainable exploitation of resources are affecting the well-being of our planet, is becoming more and more evident in our time. What the full result of this will be remains to be seen by future generations.

Should we, for example, discover ways in which we could manufacture water and oxygen, our options would increase dramatically. We are however not able to create *per se*, so if we insist on continuing our present lifestyle, additional resources would have to be found elsewhere in the universe. Therefore, I believe it is unrealistic to wipe the concept of a pending "Doomsday" completely off the table.

By the way, I believe that extreme care should be taken when we attempt to understand the messages of the authors who wrote the apocalyptical literature found in the biblical books of Daniel and Revelations. Both these books were written to specific target populations experiencing specific traumatic circumstances as a nation and as individuals at that time. Because there are many of us who are in very much the same situation as the original audiences, these books remain relevant, but only as a source of encouragement under trying circumstances and not as an instrument to fish out details about our future here on earth and into eternity. Also, I disagree with those who believe that the Jewish nation as a physical entity will have a determining influence on world affairs in future. I believe that in the post-Jesus era, God deals with humanity as a whole. Although I believe in the preservation of nations, cultures, and languages, I also believe that the globalization process will force us to acknowledge that our destiny cannot and will not be determined on those basses alone. Apartheid, for example, is a political system that does not acknowledge the above, and could therefore not succeed in South-Africa, and will not succeed in any one of the many other places in the world where it is practiced today. The preservation of our natural heritage has to be dealt with on a global scale. In the Middle East, for example, water is gradually surpassing political ideologies as an important factor in the survival of a number of nations, and a growing number of decision makers are realizing that the preservation of the bird migratory routes is crucial to the region. The struggle in our era is between "Good" and "Evil" in the widest sense of these words. (See par. 3.2.1 and par. 6.4)

In the mean time, I am convinced that God expects us to use our intelligence and the resources at our disposal, in a global stewardship, to slow down these destructive processes. Fortunately there are those amongst us that are making a difference. I consider myself privileged to know something about one such person. His name? John Denver.

e) John Denver

Henry John Deutschendorf (1943–1997), is better known as the folk singer/songwriter, John Denver. He took his professional surname from the Rocky Mountain city that he loved so much, sung about, and later made his home. His albums earned fourteen gold and eight platinum awards in the USA alone. His album *John Denver's Greatest Hits* is still one of the best selling albums in the history of RCA Records, with international sales having passed ten million copies recently.

It is said the album ignited a popular consciousness that celebrated the wonders of nature in many hearts and minds in the USA and even globally. That, because of the massive appeal of the album, it significantly helped to popularize the conservation ethic in not only the American but also in many other cultures around the globe. To this day, he is one of the top five recording artists in the history of the music industry. He had been dubbed "the poet for the planet", because so much of his life and many of his songs reflect his love and concern for the environment.

In life, he was a co-founder of the Windstar Foundation, which exists to this day. It is an environmental organization which promotes a holistic approach to global problems. This is done mainly by educating its members all over the world to make a difference at the local level rather than embarking on grandiose projects. In addition, he was a co-founder of Plant-it 2000, an organization dedicated to urging people worldwide to replenish destroyed flora, and have planted 250 000 indigenous trees in five years. John served on the Board of Advisors of the Wildlife Conservation Society, assisting in the establishment of more than one hundred parks and reserves around the world, including the Arctic National Wildlife Refuge in Alaska. He served as a member of the American Presidential Commission on World and Domestic Hunger. For his dedication to and work in aid of the global environment, he received, amongst others, the following awards:

- The Presidential "World Without Hunger" Award.
- The American Jaycees "Ten Outstanding Men of America" Award.
- The "Whale Protection Fund Service" Award.
- The NASA "Medal for Public Achievement" Award.
- The International Centre for Tropical Ecology's "World Ecology" Award.
- The "Albert Schweitzer Music Award" in 1993 for: "a life's work dedicated to music and devoted to humanity." It was the first time ever, that a non-classical musician had received this honor.

Why all this about a recording artist in a book dealing with *Birds and Bibles in History*? Why not rather Sir David Attenborough, Bill Oddie, Al Gore, Greenpeace, or Captain Jacques-Yves Cousteau? The efforts of the latter in educating people by distributing his research findings on the importance of the ocean through television programs are well known. Incidentally, he was a personal friend of John Denver. They spent some time together on Cousteau's well-known research ship *Calypso* and John wrote a song about the ship and the work that was done on it. John went a step further and donated all income generated by the song to the Cousteau foundation. Let John Denver answer the

John Denver
(Photo: Mary Ledford) [9]

above question himself through the words of a song called *Boy from the Country*. It was written by Michael Murphey and often performed by John. It says that the Boy from the Country tried to tell us that the animals could speak, and then poses two questions: Who knows, perhaps they do? How do you know they don't, just because they haven't spoken to you?

I believe that the words of the song summarize the message that John Denver ultimately wanted to convey to the world. On his album *Aerie*, he pleads for the birds and the environment through the song *The Eagle and the Hawk* by saying that he is the eagle, living in high country, in rocky cathedrals that reach to the sky. He is the hawk and there's blood on his feathers. But time is still turning, they soon will be dry. And then he concludes by saying that all those who see him and all who believe in him, share in the freedom he feels when he flies.

Birders all around the world would readily agree that the birds do actually speak to us, maybe not vocally, but surely through their splendor, their habits, and their talents. This, I believe, is why biblical authors, and even Jesus Christ himself, from time to time looked at the birds and employed them to illustrate their teachings in a manner that made it understandable to the most modest of layman. Could this, thousands of years later, be one of the reasons for the phenomenal growth in the number of people taking up birding as a hobby?

John Denver's other passion was flying. He was an accomplished aerobatic pilot, had flown American Air Force F–15 fighters, and the space shuttle simulator. In 1988 he applied, unsuccessfully, to the USSR for permission to become the first civilian to visit the space station Mir. He died flying, when the experimental plane that he was piloting, crashed into Monterey Bay, off the coast of Pacific Grove, California. Ironically, one of his favorite own-compositions was *The Wings That Fly Us Home*, and it was with words from

---

[9]   Photograph of John Denver performing by Mary Ledford (EaglesHorses Productions). The photograph was taken in Birmingham, Alabama, in February of 1997. Used with permission.

this song that he ended his 1994 autobiography: "Yesterday I had a dream about dying, about laying to rest and then flying." He then goes on to say: "How the moment at hand, is the only thing we really own. And I lay on my bed and I wonder, after all has been said and done. Why is it thus we are here, and soon we are gone…"

Rüppell's Vulture *(Gyps rueppellii)*
(Photo: Tony Hisgett) [10]

---

[10]   This image is a reproduction of the photograph taken by Tony Hisgett in Moreton in Marsh, England and used with permission. Originally placed on www.flickr.com, the original photograph can be viewed at: http://www.flickr.com/photos/hisgett/6022576576/.

DEDICATION

This book is dedicated to my Grandfather:
Christiaan Johannes Hattingh
(1894–1965)

Who quietly cried,
when the Malachite Sunbird in his garden, died.

Grandpa Christiaan and I. *(ca. 1954)*
(Photo: Tian Hattingh) [11]

Malachite Sunbird *(Nectarinia famosa)*
(Photo: Alan Manson) [12]

---

[11] This picture is from my own collection.

[12] This image is a reproduction of the photograph taken by Alan Manson and used with permission. Originally placed on www.flickr.com, the original photograph can be viewed at: http://www.flickr.com/photos/12457947@N07/3281380084/in/photostream.

BIRDS AND BIBLES
IN HISTORY

PART I
ORNITHOLOGICAL HISTORY

---

*An overview of a number of the personalities and
events in the ornithological world,
set in the general history of mankind
from ancient to modern times.*

Namaqua Dove *(Oena capensis)*
(Photo: Dr. Marcel Holyoak) [1]

CHAPTER 1
ORNITHOLOGICAL HISTORY

> "In the beginning God created the heavens and the earth. And God said: Let birds fly above the earth, across the expanse of the sky. And God created... every winged bird according to its kind. God blessed them and said: Be fruitful, and let the birds increase on the earth. And there was evening, and there was morning, the fifth day."
>
> **Gen 1:1, 20–23 (NIV)** [2]
>
> ---
>
> **ABC Online: May 31, 2010**
>
> Scientists say an Aboriginal rock art depiction of an extinct giant bird could be Australia's oldest painting. "Initially, we thought it was another big emu," said consulting archaeologist Ben Gunn, a founding member of the Australian Rock Art Research Association. "The animal wasn't an emu; it looked like the megafauna bird Genyornis, an emu-like, big-beaked, thick-legged bird that went extinct along with other Australian megafauna between 40,000 and 50,000 years ago," he stated. [3]

1.1 In the Beginning

Ornithology, noun: from the Greek: *ornithos*, meaning "bird"; and *logos*, meaning "rationale" or "explanation", is a branch of the science of zoology that is exclusively concerned with all aspects of the study of birds.

As it will be evident from this publication as well, the science of Ornithology has a long and interesting history. The American zoologist G.M. Allen (1879–1942) rightly pointed out that the study of birds is by and large an auspicious (of good omen, favorable; prosperous) activity. In this regard we should keep in mind that the word auspicious is derived from two Latin words, *aves* meaning "bird" and *spicere* meaning "to see." In ancient times (and to some extent today), the custom of forecasting the outcome of im-

---

[2] THE HOLY BIBLE, NEW INTERNATIONAL VERSION®, NIV® Copyright © 1973, 1978, 1984, 2011 by Biblica, Inc.™ Used by permission. All rights reserved worldwide.

[3] Reproduced by permission of the Australian Broadcasting Corporation (c) 2010 ABC. All rights reserved.

portant events and/or results of undertakings by observing the flights and/or calls of birds or even by examining their entrails and interpreting the findings, led to the formation of the word.

Although Ornithology over the centuries has advanced into a fully-fledged science, as will be shown in the next paragraphs, it is still regarded as an auspicious activity to this day. The main reason being that the destinies of both man and the avian kingdom (Class: *Aves)* have always been, and will always be, closely linked. To employ the status of the resident bird population in any ecosystem as an indicator of the overall condition of such a system has become a common scientific practice in our day.

Birds are not only highly attractive and therefore visible animals, but also very common in our world. The newsflash above serves as an example confirming that birds have always had an aesthetic appeal to us humans since the beginning of time, resulting in a very special relationship with man. Birds, as observed by pre-historic man, are without exception featured in rock art and rock paintings all over the globe. However, this relationship was established initially, and maintained through the ages, for another even more important reason.

A midden, also known as a kitchen midden, or a shell heap or shell mound, is a dumpsite created by humans for domestic waste disposal purposes. Archaeologists worldwide are interested in them as they contain waste products relating directly to day-to-day human life. They contain a detailed record of what food was processed and/or eaten by the people who initially created them. Eggs laid by females of many different animal species, including birds, reptiles, amphibians, and fish, have probably always been favoured and consumed by mankind. Wild birds as well as their eggs have been sources of food (specifically high protein) for humans for millennia. So it is not surprising that water bird and seabird remains have been found in shell mounds (middens) on the island of Oronsay off the coast of Scotland. In a Stone Age hut in Israel, bones of over 80 species of birds have been found.

The Vedas (From Sanskrit, meaning "knowledge") are a large body of texts originating in ancient India. Composed in Vedic Sanskrit, the texts constitute the oldest layer of Sanskrit literature and the oldest scriptures of Hinduism known to us, and are dated roughly 1500–1000 BC. The Vedas mention the habit of brood parasitism in the Asian Koel (*Eudynamys scolopacea*), a member of the Cuckoo order of birds, the *Cuculiformes*, which is found in South Asia, China, and South-east Asia.

It is generally accepted that Bibles should always be regarded as a major contributor in any research, including this one, where the ancient history of the Middle East is involved. However, I also believe that one should always keep in mind that ancient Israel was a buffer state between her two great neighbors, Egypt to the south, and Mesopotamia to the north-east. Therefore, I am convinced that Biblical evidence should be treated in the context in which it evolved. In my research I therefore had to be prepared to gather information from a wide variety of sources, in order to confirm facts as far as possible, and decrease the possibility of a biased interpretation of history.

From the outset, and with the above in mind, I have to make a further qualification, which, in the framework of the objectives of this book, would ease our task tremendously, without jeopardizing the objectives thereof. I believe that if all that has been written on the dating of the events and personalities described in the Old Testament, as well as the dating of the individual books themselves, had to be read by one single person, s/he would need more than one lifetime to do so. Therefore, for the purposes of this book, I shall make a number of presumptions regarding Biblical and other dates, based on popular, and in my humble judgment, acceptable views. We simply cannot do better than that.

The events described in Bibles, were never meant to be a scientifically accurate description of what actually happened. This is even more so the case regarding Genesis Chapter 1–11. Descriptions of the creation of the world, of a great flood, and so on, were well known in neighboring as well as far-off countries at the time when the Biblical versions were formed. Similarities as well as differences occur between the different renderings. For example: a Sumerian flood myth dating back to 6000 BC, agrees with the Hebraic version (Gen 6–8), in mentioning that a "dove" was sent out by Noah to test the level of the receding waters. The similarities do, however, not in any way imply compatibility in outlook and basic belief. Even as far away as Australia, aboriginal folklore include a large flood resulting from the cruel treatment of *Dumbi the Owl* by some children. Let us assume that these events took place somewhere before, and up to the year 2000 BC.

Following Genesis 1–11, we find the history of the Patriarchs (Gen 12–50) that are placed in 2000–1650 BC; the Egyptian slavery and Exodus, taking up 1650–1240 BC; Saul (ca. 1079–1007 BC), David (ca. 1011–971 BC), and Solomon (ca. 971–931 BC). In 722 BC the Kingdom of Israel fell to the Assyrians, and in 587 BC Judah fell to the Babylonians. The exile (587–539 BC) disrupted the Jewish life in Palestine until the *Edict of Restoration* by the Persian ruler Cyrus, resulting in the rebuilding of the Temple by March 515 BC, and the walls of Jerusalem by December 437 BC.

Alexander the Great (356–323 BC) was the king of Macedon from 336 BC, and defeated the Persians in 333 BC, then turned southwards along the Mediterranean coast, through Palestine, and conquered Egypt by 332 BC. After his death his empire fell apart and his general, Ptolemy gained control of Palestine. He in turn was defeated by the Seleucid, Antiochus in 198 BC. The Maccabean War 166–164 BC, was followed by the Hasmonian rule (155–65 BC), followed in turn by the Roman rule (65

BC–320 AD). Jesus Christ (3 BC–30 AD) lived and worked in Palestine, and founded Christianity.

The second Temple was destroyed by the Romans in 70 AD, and most of the Jewish people were scattered to many parts of the world, the so-called *Diaspora*. Late in the nineteenth century they started returning to Palestine in significant numbers, and after living in the *Diaspora* for 1878 years, on 15 May 1948 AD, the political state of Israel was proclaimed by David ben Gurion.

1.2 The Old Testament Authors

There are very few facts known about the dating and the authors of the books of the Old Testament, partly as a result of the custom to destroy the original Hebrew texts after they had been copied. This was done to prevent the desecration of texts by enemies and/or foes. As set out in the concluding statement (See par. 27), I am convinced that the authors and/or editors of the books of the Old Testament, have to be given serious consideration by anyone seeking for the origins of Ornithology and/or the original ornithologists. In the context of this book we are briefly discussing only four of them, chosen because of the relatively high occurrence of birds in their writings.

1.2.1 Moses

It is generally accepted that Moses received a thorough education while in Egypt. There is no concrete evidence that he wrote the Pentateuch, but personally I would accept the view that he could be seen as an editor and/or compiler. He may have received direct revelations from God, (e.g. Ex 3; 20; 33:11), but I feel more at ease with the view that he faithfully set down orally transmitted records that came to him from past generations. Copyists might have added "footnotes" (e.g. Gen 12:6) up to the time of the Monarchy, but they would not constitute major changes.

1.2.2 Job

Containing some of the most difficult poetry in the Old Testament, and a vocabulary with some one hundred and ten words not found elsewhere, it is an extremely difficult text to study and translate. Scholars have been of the opinion that the poet ranks with the greatest of mankind. They further agree that its creative genius is so original that it does not fit into any of the standard categories devised by literary criticism. The book is anonymous, and no consensus had yet been reached on its date. The absence of any of the details of the Israelite history suggests an early date. The mention of the Chaldeans as nomadic raiders (1:17), points to the antiquity of the story, possibly as early as Patriarchal times. Dates between 600 and 400 BC are the most popular, with a Solomonic date (ca. 971–931 BC) the earliest we could possibly adopt. For the purpose of the "Conclusions" given in Chapter 27, the reader should keep in mind that Aristotle lived from 384 BC to 322 BC.

1.2.3 Isaiah

It is now generally accepted that the book is a compilation of the prophesies collected from many sources (e.g. 1:1, 2:1, 13:1), brought together into one volume. All of these sources came into being as the result of a long and gradual process, which started with oral traditions transferred from one generation to the next. The book is divided into three sections namely:

- Chapters 1–35, compiled by the so-called "first" Isaiah, containing prophesies against Judah and Jerusalem (1–12), other heathen nations (13–23), apocalyptic prophesies (24–27), and against the people of the Covenant (28–35). This section is dated somewhere in the eighth century BC;
- Chapters 36–39 forms a historical interlude by the "second" or Deutero-Isaiah;
- Chapters 40–66, with chapters 40–55 by the "second" Isaiah, dated in the time between the destruction of Jerusalem in 587 BC and the downfall of the Babylonian empire in 539 BC, and chapters 56–66 by the "third" or "Trito-Isaiah", from the post-exilic period.

1.2.4 Jeremiah

Born in ca. 650 BC, Jeremiah worked under the last five kings of Judah, during the forty years 626–587 BC. He was a man of marked contrasts, being at once gentle and tenacious, affectionate and inflexible. In Jer 36:4 we find Baruch mentioned as the prophets' secretary, recording his masters words in writing and reading them in public (36:6) on behalf of Jeremiah. The book is divided into parts according to the king in whose time the prophet had been active then. For example: Chapters 1–6 under King Josiah, and Chapters 7–20 under King Jehoiakim.

1.3 Aristotle

Socrates (469–399 BC) had two famous students: Plato (424–348 BC) and Xenophon (ca. 430–354 BC). Xenophon of Athens was a Greek historian, soldier, mercenary, and a contemporary and admirer of Socrates. He for example recorded the abundance of the Ostrich *(Struthio camelus syriacus)* in Assyria (In Anabasis, i. 5).

Aristotle (384–322 BC), the most famous of the Greek philosophers, was a student and disciple of Plato. He remained at Plato's Academy for nearly twenty years before leaving in 348/47 BC after Plato's death. Later he undertook the education of Alexander the Great in 343 BC.

Wikipedia, in its article on "Aristotle" contains the following statement to illustrate the immensity and scope of his work. "Aristotle not only studied almost every subject possible at the time, but made significant contributions to most of them. In physical science, Aristotle studied Anatomy, Astronomy, Embryology, Geography, Geology, Meteorology, Physics, and Zoology (including Ornithology). In Philosophy, he wrote on Aesthetics, Ethics, Government, Metaphysics, Politics, Economics, Psychology, Rhetoric, and Theology. He also studied Education, Foreign customs, Literature, and Poetry. His combined works constitute a virtual encyclopedia of Greek knowledge. It has been suggested that Aristotle was probably the last person to know everything there was to be known in his time." [4]

He compiled the first known scientific list of birds, containing one hundred and seventy species. In his encyclopedia of zoology, *Historia Animalium,* consisting of nine volumes, he studied animals, fish and birds objectively. His descriptions of fish anatomy were unsurpassed for two thousand years, until John Ray studied the subject. (See par. 2.2). He studied the biology and behavior of the octopuses, squid and paper argonauts of the Mediterranean. He observed how woodpeckers *(Picidae),* placed nuts in crevices to crack the shells more easily, and mentions the antagonistic relationship between owls and crows.

What did he actually write? For example, about the bathing habits of birds he said that some take a dust-bath by rolling in dust, some take a water-bath, and some take neither the one bath nor the other. He goes on to say that birds that do not fly but keep on the ground take the dust-bath, as for instance the hen, the partridge, the francolin, the crested lark, and the pheasant. Some of the straight-taloned birds, and such as those that live on the banks of a river, in marshes, or by the sea, take a water-bath. Some birds take both the dust-bath and the waterbath, as for instance the pigeon and the sparrow. Some

---

4   Wikipedia contributors. "Aristotle." Wikipedia, The Free Encyclopedia. Wikipedia, The Free Encyclopedia, 26 Jul. 2011. Web. 6 Aug. 2011.

Aristotle (384–322 BC)
(Photo: Wikimedia Commons) [5]

of the crooked-taloned birds take neither the one nor the other.

Aristotle observed, but could not explain, the aggressive rivalry amongst male birds in springtime. In his encyclopedia he searches for a "scheme of nature", employing a "ladder of life", attempting to illustrate the relationships between different living creatures, and appreciating that there are great divisions in the animal kingdom. In hindsight, we might find it strange that he classified animals according to their size, so that the Elephant *(Proboscidae)* was the first mammal and the Ostrich *(Struthio camelus)* the first bird, but we should keep in mind that he had absolutely no references to consult or beacons to work from.

Swallows *(Hirundinidae)* drink in flight, by swooping in over the water, placing their beaks into the water, and scooping up as much as needed. They also have the habit of roosting in reed beds. These two facts led people like Aristotle to fervently believe that swallows hibernated underwater during the winter. Refer to par. 10.5 on what Aristotle had to say about the Indian Peafowl *(Pavo cristatus)*. His work could have served as a solid foundation for biological science, but unfortunately, his spirit of free enquiry languished, to come alive again 2000 years later, at the end of the Middle Ages.

> "One swallow does not make a summer"
> **Aristotle (384–322 BC)**

---

[5] This image is a file from the Wikimedia Commons. Full details of the file can be found at: http://en.wikipedia.org/wiki/File:Aristotle_Altemps_Inv8575.jpg. Jastrow, a Wikipedia Commons user, and the copyright holder of this work, released this work into the public domain on 11 November 2006. This applies worldwide. In some countries this may not be legally possible; if so: he grants anyone the right to use this work for any purpose, without any conditions, unless such conditions are required by law.

1.4 A Chinese Encyclopedia

Although Ornithology as a science did not start until after the Opium Wars (1839–1842 and 1856–1860) resulted in the opening up of ports on the Chinese mainland, the interest in birds by the Chinese goes back for millennia. In the Shang Dynasty (1523–1100 BC) birds were commonly used as decorative motifs on utensils like bronze vessels, and as free standing finials on vessel lids. During the succeeding Chou Dynasty, from about 1100 BC, finely crafted birds are found in bronze and/or pottery. The craftsmanship was of such a high standard, that a species like the Mandarin Duck (*Aix galericulata*), can often be recognized. It was however not until the Han Dynasty (206 BC–220 AD) that paintings of birds like owls and crows first appeared.

In China the *Erya* has been described as a dictionary, glossary, thesaurus, and encyclopedia. Bernhard Karlgren explains that the book is not a dictionary *in abstracto*, but that it is a collection of direct glosses to concrete passages in ancient texts. He concluded that the major part of its glosses (explanations) must reasonably date from the third century BC. The book's author is unknown. The received text contains 2,094 entries, covering about 4,300 words, and a total of 13,113 characters. It is divided into nineteen sections. The last seven sections deal with grasses, trees, insects, reptiles, fish, birds, wild animals, and domestic animals. It describes more than five hundred and ninety kinds of flora and fauna. Chapter 17 deals with one hundred and six birds. For example: the first entry *zhui qi* means "short tailed birds" and might refer to the Rails (*Rallus Spp.*) and/or Crakes (*Porzana Spp.*). The last entry is *cang geng* and by all accounts refers to the Black-naped Oriole (*Oriolus chinensis)*. It is a valuable document of natural history and historical biogeography.

1.5 Pliny the Elder

Gaius Plinius Secundus (23–79 AD), better known as Pliny the Elder, (Pliny the Younger was his nephew), was a Roman scholar, encyclopedist, author, naturalist, and natural philosopher. He served as a naval and army commander in the early Roman Empire, and was a personal friend of the emperor Vespasian. Despite his active public life, Pliny the Elder still found time to write enormous amounts of material. He was the author of at least seventy-five books, not to mention another one hundred and sixty volumes of unpublished notebooks. His *Naturalis Historia* is one of the largest single works to have survived from the Roman Empire to the modern day and purports to cover the entire field of ancient knowledge, based on the best authorities available at the time. In fact, a novel feature of the *Naturalis Historia* was the care taken by Pliny in naming his sources. It encompasses

Pliny the Elder (23–79 AD)
(Photo: Wikimedia Commons) [6]

the fields of botany, zoology, astronomy, geology and mineralogy as well as the exploitation of those resources.

He substantially expanded Aristotle's work by adding many tall stories and amazing superstitions. For example: the Common Barn-owl *(Tyto alba)*, frequently referred to as the "Screech Owl", was portrayed as a bird of ill omen and even as an associate of witches. The latter probably stemming from its uncannily "human" face. However, there were some positive prescriptions by him as well. A few examples: The feet of the owl, burnt with the herb plumbago, was claimed to be effective against snakes. Should the heart of the owl be laid on the left hand side of a sleeping woman, she is said to disclose all the secrets of her heart. To cure hemorrhage "the bird known as the Screech Owl is boiled in oil, ewe milk butter and honey being added to the preparation when properly dissolved." Pliny's book was widely copied, and in turn formed the basis of the *Physiologus,* (natural history), which was probably written in Alexandria in about 200 AD, and became the standard reference work on nature at this point in time. The explanation of animal behavior in the "Dark Ages" was to a large extent formed in ignorance, and gave birth to many a superstition.

1.6 The Middle Ages

Medieval, adjective: Latin: *medium aevum*, meaning "the middle age."
1. Relating or belonging to the Middle Ages.
2. *Informal*: Old-fashioned; unenlightened.

---

[6] This image is a file from the Wikimedia Commons. Full details of the file can be found at: http://en.wikipedia.org/wiki/File:Plinyelder.jpg. This image is a work of the National Institutes of Health, part of the United States Department of Health and Human Services. As a work of the U.S. federal government, the image is in the public domain.

The "Dark Ages" were seen as "dark" compared to the "light" of classical antiquity. However, since the early 20th century historians prefer the term "Middle Ages." It is the middle period in a three-period division of history: Classic, Medieval, and Modern. This medieval period refers to the period of cultural and economic deterioration and disruption that occurred in Europe following the Fall of the Western Roman Empire in 476, to Flavius Odoacer (433–493), and stretching to the fall of the Byzantine (Eastern Roman) Empire in 1453 with the siege and fall of Constantinople (now Istanbul). The following subdivisions are often used:

- The Early Middle Ages   (476–1000)
- The High Middle Ages   (1000–1300)
- The Late Middle Ages   (1300–1453)

1.6.1 The Early Middle Ages (476–1000)

Christians felt obliged to acquaint themselves with God's creation, and the church became interested in the natural world. In the eleventh century monks, instead of studying nature first hand, studied the Greek and Roman works mentioned above, took note of animals described by explorers, added exquisite illustrations, and called their works *Bestiaries*. Often, apparent impossibilities were observed by the travelers, e.g.: a bird carrying a bag beneath its beak, led to the formation of some fabulous creatures. On other occasions the monks, albeit unwittingly, accurately depicted animal and bird behavior by their charming cameos e.g.: a White Stork *(Ciconia ciconia)* seizing a frog, an Osprey *(Pandion haliaetus)* carrying a fish head first (Refer to par. 12.1), Magpies *(Pica pica)* mobbing a Eurasian Eagle-owl *(Bubo bubo),* and an Indian Peafowl *(Pavo cristatus),* portrayed in its entire splendor. In the *Bestiaries* these armchair naturalists attempted to explain the relationship between nature, man, and God in terms of Christian morality, instead of studying nature *per se.*

Muhammad ibn 'Abdullah, also spelled Muhammed or Mohammed (570–632), the founder of Islam, was born in Mecca, Arabia. He fled to Medina in 622, but returned to Mecca as an acknowledged conqueror in 630.

Abu 'Ali al-Husayn ibn 'Abd Allah ibn Sina (ca. 980–1037), commonly known by his Latinized name Avicenna, was a Persian physician and philosopher. He was also an astronomer, chemist, geologist, Hafiz (a term used by Muslims for people who have completely memorized the Qur'an), Islamic psychologist, Islamic scholar, Islamic theologian, logician, paleontologist, mathematician, Maktab teacher, physicist, poet, and scientist. He is regarded as the most famous and influential individual of the Islamic Golden Age.

He published his *Abbreviatio de Animalibus*, as homage to Aristotle. At this time the Masoretes, based primarily in present-day Israel in the cities of Tiberias and Jerusalem, as well as in Iraq (Babylonia), started their work of adding vowel pointings to the consonantal text of the Old Testament. (See par. 5.3 and par. 5.4.4).

1.6.2 The High Middle Ages (1000–1300)

During this time there was one noteworthy ornithological occurrence. Frederick II (1194–1250) was one of the most powerful Holy Roman Emperors of the Middle Ages and an enlightened and educated monarch in Europe. He was interested in mathematics, architecture and natural history, and founded the University of Naples Federico II in 1224. He studied birds in general, using the information to the benefit of the sport of falconry, which was his great passion. (Refer to par. 11.2). He recorded his observations and speculations in an immense and profusely illustrated book, which he called *De Arte Venandi Gum Avibus*, "On the Art of Hunting with Birds", completing it in about 1250. The document shows the depth of his understanding in bird behavior, for example: recognizing and exploiting an important principle of animal behavior, namely that animals often react only to a narrow band of features in their surroundings. A principle that Konrad Lorenz would confirm seven hundred years later in his research in Austria on the courting displays of drakes and the reaction of goslings to different forms soaring overhead. (Refer to par. 3.5).

Frederick II (1194–1250)
(Photo: Wikimedia Commons) [7]

Frederick II categorically refuted the notion by Aristotle that birds hibernate on

---

[7] This image is a file from the Wikimedia Commons. Full details of the file can be found at: http://en.wikipedia.org/wiki/File:Frederick_II_and_eagle.jpg. This image (or other media file) is in the public domain because its copyright has expired. This applies to Australia, the European Union and those countries with a copyright term of life of the author plus 70 years.

the grounds that he observed many species making two excursions per year. He saw them flying south in the autumn, and north during springtime. He went one step further and recognized the migratory urge to apparently be of a compulsive nature. His investigations covered a wide range of ornithological fields. He understood the function of the oil gland of water birds, and the brood-parasitic behavior of many true Cuckoos (*Cuculidae*), e.g. the Common Cuckoo (*Cuculus canorus*), behavior which was shrouded in controversy for another seven hundred and fifty years. He investigated and proved as incorrect the belief that Barnacle Geese (*Branta leucopsis*) were generated from a shellfish clinging to underwater timber, and not an egg, and studied the different flight styles of birds, and meticulously recorded them.

Michael Scot (1175–1232?) was a medieval mathematician and scholar. Scot began his scholarly career as a translator. Frederick II attracted him with many other savants (learned scholars) to his brilliant court. At the instigation of the emperor he superintended a fresh translation of Aristotle and the Arabian commentaries from Arabic into Latin in 1220.

Further east, Genghis Khan (1162–1227), after much warfare, destruction, pillage and subjugation, became Khan of the united Mongol and Tartar tribes. Before Genghis Khan died, he assigned Ögedei Khan as his successor and split his empire into khanates among his sons and grandsons. He died in 1227 after defeating the Western Xia. He was buried in an unmarked grave somewhere in Mongolia at an unknown location. A few years later Marco Polo (1256–1323), the Venetian traveler and explorer, visited China and India, witnessing many wonders and marvels unknown to Europe.

Albertus Magnus, O.P. (1206–1280), also known as Albert the Great and Albert of Cologne, was a Dominican friar and a bishop, who achieved fame for his comprehensive knowledge of and advocacy for the peaceful coexistence of science and religion. His *De Avibus* was printed in 1478, which mentions many bird names for the first time.

1.6.3 The Late Middle Ages (1300–1453)

It was in the latter part of the Middle Ages, that the German, Johann Gutenberg (ca. 1398–1468) invented printing with moveable type. Gutenberg was the first European to use movable type printing, in around 1439, and the global inventor of the printing press. Among his many contributions to printing are: the invention of a process for mass-producing movable type; the use of oil-based ink; and the use of a wooden printing press similar to the agricultural screw presses of the period. His Latin Bible of 1455 is considered the first book printed wholly in this manner. By that time Frederick's work had to a large extent gone to waste, making him an isolated phenomenon in an otherwise unproductive age. However, the advent of the printing process, together with the discovery and fearless publication of the theory by Nicolaus Copernicus (ca. 1473–1543), that the sun and not the earth was at the center of the solar system, assisted in launching Europe into the Renaissance.

Johann Gutenberg (ca. 1398–1468)
(Photo: Wikimedia Commons) [8]

CHAPTER 2
THE RENAISSANCE

> "On the eighth day he must take two doves or
> two young pigeons and come before the LORD"
>
> **Lev 15:14 (NIV)** [1]

2.1 Contemporaries

Renaissance, noun: regeneration or new birth. The revival or rejuvenation of anything which has long been in disuse, decay, or extinct; the transitional movement in Europe from the Middle Ages to the so-called Modern World.

The Renaissance (French: from *ri-* "again" and *nascere* "birth"), was a cultural and mindset movement in the widest sense of the word, that spanned roughly from the late fourteenth to the late seventeenth century. It began in Florence in the Late Middle Ages and later spread throughout the rest of Europe. More and more thinking people became more and more skeptical about the present situation, and desired to free themselves from the current medieval prejudices. Consequently they bravely set out to conquer new worlds, some physical and some of the mind. As we take note of some of these men, we shall see that the Renaissance permeated all aspects of life. These men were the contemporaries of those who would bring Ornithology too, into the modern world.

2.1.1 The Fourteenth Century

Johann von Wonneck Caub, who Latinized his name to Johannes de Cuba, wrote the first book of natural history printed in that century. Little is known about him. He probably was a doctor in Frankfurt am Main. His book first appeared in German under the title of *Gart der Gesundheit* (1485) and is then translated into Latin under the title of *Hortus*

---

sanitatis (1491) and was published by Jacob von Meydenbach. *Hortus sanitatis* is divided into several treaties.

2.1.2 The Fifteenth Century

Polymath, noun: (Greek, *polymathes*, meaning "having learned much"), is a person who is very knowledgeable and whose expertise spans over a significant number of different subject areas. By today's standards, most of the ancient scientists were polymaths.

Christopher Columbus (ca. 1446–1506), the Italian navigator, discovered the Bahamas, Cuba, and other West Indian Islands.

The Portuguese navigators, Vasco da Gama (ca. 1460–1524), who discovered the sea route to India in 1498, by rounding the Cape of Good Hope, and Ferdinand Magellan (ca. 1480–1521), who's surviving crew completed the circumnavigation of the world in 1522.

Hernando Cortez (ca. 1485–1547), was a Spanish *Conquistador* who led an expedition that caused the fall of the Aztec Empire and brought large portions of mainland Mexico under the rule of the King of Castile in the early 16th century.

Leonardo da Vinci (ca. 1452–1519), "can be regarded as one of the greatest of versatile geniuses the world has ever known. He was an Italian Renaissance polymath: painter, sculptor, architect, musician, scientist, mathematician, engineer, inventor, anatomist, geologist, cartographer, botanist and writer. Leonardo has often been described as the archetype of the Renaissance Man. He has been described as man of "unquenchable curiosity" and "feverishly inventive imagination." He is widely considered to be one of the greatest painters of all time and perhaps the most diversely talented person ever to have lived. Leonardo was and is renowned

Leonardo da Vinci (ca. 1452–1519)
(Photo: Wikimedia Commons) [2]

---

[2]   This image is a file from the Wikimedia Commons. Full details of the file can be found at: http://en.wikipedia.org/wiki/File:Leonardo_self.jpg. This work is in the public domain in the United States, and those countries with a copyright term of life of the author plus 100 years or fewer.

primarily as a painter. Among his works, the *Mona Lisa* is the most famous and most parodied portrait and *The Last Supper* the most reproduced religious painting of all time, with their fame approached only by Michelangelo's *Creation of Adam.* Leonardo's drawing of the *Vitruvian Man* is also regarded as a cultural icon, being reproduced on items as varied as the euro, textbooks, and T-shirts. Perhaps fifteen of his paintings survived the small number due to his constant, and frequently disastrous, experimentation with new techniques, and his chronic procrastination." [3]

Michelangelo Buonarroti (ca. 1475–1564)
(Photo: Wikimedia Commons) [4]

The Italian Michelangelo Buonarroti (ca. 1475–1564), "was a Renaissance painter, sculptor, architect, poet, and engineer. Despite making few forays beyond the arts, his versatility in the disciplines he took up was of such a high order that he is often considered a contender for the title of the archetypal Renaissance Man, along with fellow Italian Leonardo da Vinci. Michelangelo's output in every field during his long life was prodigious; when the sheer volume of correspondence, sketches, and reminiscences that survive is also taken into account; he is the best-documented artist of the 16th century. Two of his best-known works, the *Pietà* and *David,* were sculpted before he turned thirty. Despite his low opinion of painting, Michelangelo also created two of the most influential works in fresco in the history of Western art: the scenes from Genesis on the ceiling and *The Last Judgment* on the altar wall of the Sistine Chapel in Rome. Michelangelo was originally commissioned to paint the 12 Apostles against a starry sky, but lobbied for a different and more complex scheme, representing creation, the Downfall of Man and the Promise of Salvation through the prophets and Genealogy of Christ. The work is part of a larger scheme of decoration within the chapel which represents much of the doctrine of the Catholic Church.

---

3 Wikipedia contributors. "Leonardo da Vinci." Wikipedia, The Free Encyclopedia. Wikipedia, The Free Encyclopedia, 29 Jul. 2011. Web. 6 Aug. 2011.

4 This image is from a file from the Wikimedia Commons. Full details of the file can be found at: http://en.wikipedia.org/wiki/File:Michelangelo-Buonarroti1.jpg. This image (or other media file) is in the public domain because its copyright has expired. This applies to Australia, the European Union and those countries with a copyright term of life of the author plus 70 years.

The composition eventually contained over 300 figures and had at its center nine episodes from the Book of Genesis, divided into three groups: God's Creation of the Earth; God's Creation of Humankind and their fall from God's grace; and lastly, the state of Humanity as represented by Noah and his family. On the pendentives supporting the ceiling are painted twelve men and women who prophesied the coming of the Jesus.

In a demonstration of Michelangelo's unique standing, he was the first Western artist whose biography was published while he was alive. Two biographies were published of him during his lifetime; one of them, by Giorgio Vasari, proposed that he was the pinnacle of all artistic achievement since the beginning of the Renaissance, a viewpoint that continued to have currency in art history for centuries."[5] In this regard *The Agony and the Ecstasy* (1961) is a biographical novel of Michelangelo Buonarroti written by American author Irving Stone, and is in my opinion a "must read."

Martin Luther (ca. 1483–1546), "was a German priest and professor of theology who initiated the Protestant Reformation. He strongly disputed the claim that freedom from God's punishment of sin could be purchased with money. He confronted indulgence salesman Johann Tetzel with his *Ninety-Five Theses* in 1517. His refusal to retract all of his writings at the demand of Pope Leo X in 1520 and the Holy Roman Emperor Charles V at the *Diet of Worms* in 1521 resulted in his excommunication by the Pope and condemnation as an outlaw by the Emperor.

Martin Luther (ca. 1483–1546)
(Photo: Wikimedia Commons) [6]

Luther taught that salvation is not earned by good deeds but received only as a free gift of God's grace through faith in Jesus Christ as redeemer from sin. His theology challenged the authority of the Pope of the Roman Catholic Church by teaching that the Bible is the only source of divinely revealed knowledge and opposed sacerdotalism by considering all baptized Christians to be a holy priesthood. Those who identify with Luther's teachings are called Lutherans.

---

[5] Wikipedia contributors. "Michelangelo." Wikipedia, The Free Encyclopedia. Wikipedia, The Free Encyclopedia, 21 Jul. 2011. Web. 6 Aug. 2011.

[6] This image is from a file from the Wikimedia Commons. Full details of the file can be found at: http://en.wikipedia.org/wiki/File:Luther46c.jpg. This image (or other media file) is in the public domain because its copyright has expired. This applies to Australia, the European Union and those countries with a copyright term of life of the author plus 70 years.

The translation by Luther of the Bible into the language of the people (instead of Latin) made it more accessible, causing a tremendous impact on the church and on German culture. It fostered the development of a standard version of the German language, added several principles to the art of translation, and influenced the translation into English of the King James Bible."[7]

2.1.3 The Sixteenth Century

William Turner MA (1508–1568) was an English divine and reformer, a physician, and a natural historian. He studied medicine in Italy, and was a friend of the great Swiss naturalist, Conrad Gessner. He was an early herbalist and ornithologist, and it is in these fields that the most interest lies today. In 1544 he printed a commentary of the birds mentioned by Aristotle and Pliny. This is the first printed book devoted entirely to birds.

John Calvin (ca. 1509–1564), "was an influential French theologian and pastor during the Protestant Reformation. He was a principal figure in the development of the system of Christian theology later called Calvinism. Originally trained as a humanist lawyer, he broke from the Roman Catholic Church around 1530. After religious tensions provoked a violent uprising against Protestants in France, Calvin fled to Basel, Switzerland, where he published the first edition of his highly influential *Institutes of the Christian Religion* in 1536."[8]

The English dramatist and poet, William Shakespeare (ca. 1564–1616), with owls featuring prominently when he wished to emphasize the fears and superstitions of his characters, for example in *Macbeth, Henry VI,* and *A Midsummer Night's Dream,* as such symbolism was well understood by his audiences of that time.

The Italian Galileo Galilei (ca. 1564–1642), "commonly known as Galileo, was an Italian physicist, mathematician, astronomer and philosopher who played a major role in the Scientific Revolution. His achievements include improvements to the telescope and consequent astronomical observations, and support for Copernicanism. Galileo has been called the "father of modern observational astronomy", the "father of modern physics", the "father of science", and "the father of modern science." Stephen Hawking says, "Galileo, perhaps more than any other single person, was responsible for the birth of modern science."[9]

---

[7]   Wikipedia contributors. "Martin Luther." Wikipedia, The Free Encyclopedia. Wikipedia, The Free Encyclopedia, 5 Aug. 2011. Web. 6 Aug. 2011.

[8]   Wikipedia contributors. "John Calvin." Wikipedia, The Free Encyclopedia. Wikipedia, The Free Encyclopedia, 29 Jul. 2011. Web. 6 Aug. 2011.

[9]   Wikipedia contributors. "Galileo Galilei." Wikipedia, The Free Encyclopedia. Wikipedia, The Free Encyclopedia, 19 Jul. 2011. Web. 6 Aug. 2011.

Galileo Galilei (ca. 1564–1642)
(Photo: Wikimedia Commons) [10]

The English conspirator Guy Fawkes (ca. 1570–1606) who, with others, planned the Gunpowder Plot. The plan to blow up the House of Lords during the State Opening of Parliament on 5 November 1605 was thwarted, but it led to annual celebrations, the so-called "Bonfire Nights" throughout the Commonwealth.

Conrad Gessner's *Historic Animalium qui est de Auium natura* and Pierre Belon's (Bellonius) *Histoire de la nature des Oyseaux* are published in 1555. Belon lists birds according to a definite system.

In 1591, Joris Hoefnagel starts to work for Rudolf II, Holy Roman Emperor and produces for him ninety oil-based paintings, of which one is of the Dodo *(Raphus cucullatus)*.

Li ShiZhen, (1518–1593), was one of the greatest Chinese herbologists and acupuncturists in Chinese history. In 1596 his *Compendium of Chinese Materia Medica* includes a total of seventy-seven species of birds. The book was reprinted frequently, and five of the original editions still exist.

Late in this century printed versions of the *Bestiaries* became widely available, and superstition still thrived. For example: in his *Booke of Secretes* (1525), Albertus Magnus offered a different version to the advice by Pliny, by stating: "For if the heart and right foot of the Common Barn-owl *(Tyto alba)* be put upon a man sleeping, he shall say anon to thee whatsoever thou ask of him." Michelangelo, in his sculpture *Night*, depicted a Common Barn-owl standing alert at the feet of a naked, sleeping woman, representing the theme that the owl could be some sort of a guardian against the darkness. The information gained from the *Bestiaries* was incorporated into folk-tales and fables, which used animals as convenient metaphors for human behavior, with the purpose of making a strong moral statement. For example: the cunning nature of *Reynard the Fox*, from the Reynard cycle (a literary cycle of allegorical French, Dutch, English, and German fables largely concerned with Reynard, an

---

[10] This image is a file from the Wikimedia Commons: Full details of the file can be found at: http://en.wikipedia.org/wiki/File:Justus_Sustermans_-_Portrait_of_Galileo_Galilei,_1636.jpg. This image (or other media file) is in the public domain because its copyright has expired. This applies to Australia, the European Union and those countries with a copyright term of life of the author plus 70 years.

anthropomorphic red fox and trickster figure). Also the wisdom or foolishness of the owl as in *The Owl and the Farmer* from John Gay's *Fables* of 1727.

2.1.4 The Seventeenth Century

Rene Descartes (1596–1650), Rembrandt Harmensoon Van Rijn (1606–1669), Blaise Pascal (1623–1662), John Ray (1627–1705), (See par. 2.2), Sir Isaac Newton (1642–1727), widely regarded as one of the most influential people in human history, Antonio Stradivari (1644–1737), Vitus Bering (1681–1741), Johann Sebastian Bach (1685–1750), George Frederick Handel (1685–1759), and Gabriel Daniel Fahrenheit (1686–1736).

Wikipedia, in their article "Timeline of Ornithology"[11], lists no less than eleven events significant in the development of ornithology during this century. The last event being the death, in 1681, of the last Dodo *(Raphus cucullatus)* on the island of Mauritius.

Benjamin Franklin (1706–1790)
(Photo: Wikimedia Commons) [12]

2.1.5 The Eighteenth Century

John Wesley (1703–1791), was a Christian theologian and also a Church of England cleric, who, along with his brother Charles, is largely credited as the founders of the Methodist movement and denomination.

Benjamin Franklin (1706–1790), "was one of the Founding Fathers of the United States. A noted polymath, Franklin was a leading author, printer, political theorist, politician, postmaster, scientist,

---

[11] Wikipedia contributors. "Timeline of ornithology." Wikipedia, The Free Encyclopedia. Wikipedia, The Free Encyclopedia, 7 Jun. 2011. Web. 31 Aug. 2011.

[12] This image is from a file from the Wikimedia Commons. Full details of the file can be found at: http://en.wikipedia.org/wiki/ File:BenFranklinDuplessis.jpg. This work is in the public domain in the United States, and those countries with a copyright term of life of the author plus 100 years or fewer.

George Washington (1732–1799)
(Photo: Wikimedia Commons) [14]

musician, inventor, satirist, civic activist, statesman, and diplomat. As a scientist, he was a major figure in the American Enlightenment and the history of physics for his discoveries and theories regarding electricity. He invented the lightning rod, bifocals, the Franklin stove, a carriage odometer, and the glass armonica. He formed both the first public lending library in America and the first fire department." [13]

Carl von Linne (1707–1778), became one of the most distinguished of naturalists, and the founder of modern Botany, and Gilbert White (1720–1793), whose success as a naturalist stemmed from his intimate knowledge of a small patch of countryside. Their work is so influential that I discuss them in paragraphs 2.3 and 2.4 respectively.

George Washington (1732–1799) "was the dominant military and political leader of the new United States of America from 1775 to 1799. He led the American victory over Great Britain in the American Revolutionary War as commander-in-chief of the Continental Army in 1775–1783, and he presided over the writing of the Constitution in 1787. He was the unanimous choice to serve as the first President of the United States (1789–1797)." [15]

Wolfgang Amadeus Mozart (1756–1791), was a prolific (he composed over 600 works) and influential composer of the Classical era. While most composers specialize in certain kinds of pieces, Mozart created masterful works for almost every category of music - vocal music, concertos, chamber music, symphonies, sonatas, and opera. Mozart was no doubt the greatest child star that ever lived. He was traveling all over Europe playing music by the time he was six. Because of his constant travels, Mozart eventually learned

---

[13]   Wikipedia contributors. "Benjamin Franklin." Wikipedia, The Free Encyclopedia. Wikipedia, The Free Encyclopedia, 2 Aug. 2011. Web. 6 Aug. 2011.

[14]   This image is a file from the Wikimedia Commons. Full details of the file can be found at: http://en.wikipedia.org/wiki/File:Portrait_of_George_Washington.jpeg. This work is in the public domain in the United States, and those countries with a copyright term of life of the author plus 100 years or fewer.

[15]   Wikipedia contributors. "George Washington." Wikipedia, The Free Encyclopedia. Wikipedia, The Free Encyclopedia, 5 Aug. 2011. Web. 6 Aug. 2011.

to speak fifteen different languages. He is among the most enduringly popular of classical composers.

Napoleon Bonaparte (1769–1821), was a French military and political leader during the latter stages of the French Revolution, often portrayed as a power hungry conqueror. As Napoleon I, he was Emperor of the French dominating Europe for eleven years from 1804 to 1815. He is best remembered for his role in the wars led against France by a number of coalitions, the so-called Napoleonic Wars. It was as a result of these wars, and his success in them, that he is generally regarded as one of the greatest military commanders of all time, but also as a risk taking gambler. He was a workaholic genius (he slept

Ludwig van Beethoven (1770–1827)
(Photo: Wikimedia Commons) [16]

four hours a day), and an impatient short term planner. Emperor Napoleon proved to be an excellent civil administrator. One of his greatest achievements was his supervision of the revision and collection of French law into codes.

Ludwig van Beethoven (1770–1827), was a German composer and pianist, and remains one of the most famous and influential composers of all time. Apart from his other works, he only wrote nine symphonies. Beethoven   is acknowledged as one of the giants of classical music; occasionally he is referred to as one of the "three Bs" (along with Bach and Brahms) who epitomise that tradition. He was also a pivotal figure in the transition from the 18th century musical classicism to 19th century romanticism, and his influence on subsequent generations of composers was profound.

Captain James Cook (1728–1779). In three voyages Cook sailed thousands of miles across largely uncharted areas of the globe. He mapped lands from New Zealand to Hawaii in the Pacific Ocean in greater detail and on a scale not previously achieved. As he progressed on his voyages of discovery he surveyed and named features, and recorded

---

[16]   This image is a file from the Wikimedia Commons. Full details of the file can be found at: http://upload.wikimedia.org/wikipedia/commons/6/6f/Beethoven.jpg. This image (or other media file) is in the public domain because its copyright has expired. This applies to Australia, the European Union and those countries with a copyright term of life of the author plus 70 years.

islands and coastlines on European maps for the first time. He displayed a combination of seamanship, superior surveying and cartographic skills, physical courage, and an ability to lead men in adverse conditions.

He and the crew of *HMS Endeavour*, in September 1769 reached New Zealand, and in April 1770, they became the first Europeans to reach the east coast of Australia, making landfall on the shore of what is now known as Botany Bay.

Cook was killed in Hawaii in a fight with Hawaiians during his third exploratory voyage in the Pacific in 1779. He left a legacy of scientific and geographical knowledge which was to influence his successors well into the 20th century and many memorials worldwide have been dedicated to him.

It was an exceptional century in ornithological terms. Wikipedia lists no less than fifty three dates significant to ornithology in their article "Timeline of Ornithology".[17]

Clerics were educated men too, who had been steeped in the mindset of folk tales and the *Bestiaries*. As the new intellectual climate spread across Europe, they became impressed by this new process of scientific inquiry. It was therefore just a matter of time before they too began to question the information available to them. Their new discoveries proved beyond doubt that God's world were even more marvelous than they had previously suspected, and at last they undertook the study of animal behavior beyond the realm of fantasy. Let us look at a few of these men who had a profound influence on the ornithology of today.

2.2 John Ray

John Ray (1627–1705), from the village of Black Notley, in north Essex, England, and one of his students, the Hon. (later Sir) Francis Willughby traveled extensively through England and Europe, observing nature and specifically birds. He was the first to group animals and plants on the basis of their anatomical similarity, and not by the various other criteria as used in the past by Aristotle and medieval naturalists. They, for example, placed the Terns *(Sterninae)* with the Gulls *(Larinae)*, who today are subfamilies of the family *Laridae* in the Order *Charadrilformes*, and not with the Swallows *(Hirundinidae)* who are in the Order *Passeriformes*. They realized that the Blackcock and Greyhen were not two different species, but the male and female Black Grouse *(Tetrao tetrix)*.

But one of their most dramatic discoveries was that birds were territorial creatures by nature. Ray had read the book by G.P. Olina, published in Rome in 1622, in which

---

17   Wikipedia contributors. "Timeline of ornithology." Wikipedia, The Free Encyclopedia. Wikipedia, The Free Encyclopedia, 7 Jun. 2011. Web. 31 Aug. 2011.

he says that the Nightingale *(Luscinia mega-rhynchos)*, sings in its *freehold*, or estate or territory. Ray went further, and observed that the male bird "will not admit any other nightingale but its mate", into its *freehold*.

Willughby died in 1672, and Ray returned to Essex and started writing. He probably wrote most of Willughby's illustrated handbook of birds, *Ornithologia*, which was published posthumously in 1687. It was packed with information and illustrations, and basilisks, dragons and phoenixes were not to be found in it. It could safely be described as "the first scientific ornithological publication" of all times.

Ray interpreted his findings in religious terms. He saw the great variety in nature, in their physical forms as well as behavior, as a demonstration of the power of God as the Creator. For example: the adaptations animals possessed assisting them in self-preservation, the urge to multiply, why birds laid eggs rather than being viviparous (development of the embryo inside the body of the mother, eventually leading to live birth), the different designs in nests, and so on, were all "part of the Great Design of Providence." He was one of the first to write about the instinctive behavior of animals. He defined it as the behavior in an animal that is directed to ends or results unknown to the particular animal, and ascribed these phenomena to acts of God.

This culminated in his most widely read book *The Wisdom of God Manifested in the Works of Creation*, in which his real scientific genius was revealed. At the time fossils were though to be the remains of creatures drowned in the Great Flood of Genesis 6–9. He realized that the fossils pointed to creatures that were extinct, and this was very confusing to him, as, according to the church, it meant that the universe, and therefore its Creator was

John Ray (1627–1705)
(Photo: Wikimedia Commons) [18]

---

[18] This image is a file from the Wikimedia Commons. Full details of the file can be found at: http://en.wikipedia.org/wiki/File:John_Ray_from_NPG.jpg. This is a faithful photographic reproduction of an original two-dimensional work of art. The work of art itself is in the public domain. The official position taken by the Wikimedia Foundation is that this photographic reproduction is therefore also considered to be in the public domain.

Carl von Linne (1707–1778)
(Photo: Wikimedia Commons) [19]

not perfect. By the time he died more and more doubt was being cast on this point of view, but it would not be resolved for another one hundred and fifty years before Darwin's synthesis.

2.3 Carl von Linne

At this point in time, Carl von Linne (1707–1778), who changed his surname to Linnaeus, a Swedish doctor, scientist, and taxonomist, appeared on the scene. He became one of the most distinguished of naturalists, and the founder of modern Botany. His *Systema Naturae* was published in 1735, followed by other monumental works. He was the first person to expound and clear up the true principles for defining a species. To this day the principles of his system of classifying nature, as set out in the tenth edition of his work, published in 1758, are being used. In that edition, von Linne classified 4 236 species. He invented the binominal (two-name) nomenclature, by which a species is identified by the name of the genus in which it occurs, in other words its generic (general) name, coupled to the species (specific) name of the individual. (See par. 4.6.1). For a schematic classification of the living world, refer to Chapter 30, and for an example of a modern day classification of a particular species by this system, see Chapter 31.

2.4 Gilbert White

Again the church would have an influence on the development of ornithology. Gilbert White (1720–1793), was a country parson who settled in Selborne, a village to the south-

---

[19] This image is a file from the Wikimedia Commons. Full details of the file can be found at: http://en.wikipedia.org/wiki/File:Carl_von_Linn%C3%A9.jpg. This image (or other media file) is in the public domain because its copyright has expired. This applies to Australia, the European Union and those countries with a copyright term of life of the author plus 70 years.

west of London, as a curate in 1756, where he lived until his death. His success as a naturalist stemmed from his intimate knowledge of a small patch of countryside. He studied the birds, the weather and the rainfall. He described the nocturnal Bats *(Chiroptera)* and the Harvest Mouse *(Micromys minutus)*. He also studied the insects. His descriptions of the Field Cricket *(Orthoptera)* have hardly been improved since. Another testimony to the quality of his work is found in a statement made in Volume 5 of the *Handbook of the birds of the World*, published two hundred and six years after his death. The Handbook states: "The first breeding studies on the Common Barn-owl *(Tyto alba)*, date back over a century, with observations at the nest going back even farther, to Gilbert White's accurate and highly detailed remarks, on a pair breeding in his own church at Selborne, in Hampshire, England."[20]

He studied the habits of the Swallows *(Hirundinidae)* and Swifts *(Apodini)* living in the area, and noticed that, when building their nests: "Providence had endowed different members of the same tribe with such varying architectural skills." His religious convictions did not deter him from objective scientific research, and he advanced further and investigated the diet of swallows and Martins *(Hirundinidae)*, to see if their different flying habits enabled them to intercept specific varieties of insects.

He was the first person to utilize behavior in the form of song as a means of identifying bird species. For example, he was able to sort out the Leaf Warblers *(Phylloscopidae)* from the "willow wrens." He observed the difference in calls, and today the *Chirper* is called the Chiffchaff *(Phylloscopus collybita)*, and the almost identical bird with the "joyful song", the Willow Warbler *(Phylloscopus trochilus)*. He established the Grasshopper Warbler *(Locustella naevia)* and Wood Warbler *(Phylloscopus sibilatrix)* as definite species, and identified the Sedge Warbler *(Acrocephalus schoenobaenus)* from, what we know today as the Aquatic Warbler *(Acrocephalus paludicola)*.

It was the phenomenon of migration that baffled his parochial (pertaining to a parish) mind. His brother John wrote to him about migrations over Gibraltar, but the odd individual over-wintering in a numb and sluggish state, led him to believe that they retired into the banks of pools and rivers to escape the inclement winter weather. At least he had advanced on Aristotle's view, and conceded that the birds did not hibernate in the water. His objectivity and drive to find the truth, prompted him during one spring to send his gardeners into suitable areas, thrashing through the vegetation in an attempt to flush the hibernating birds out. Of course, none were found.

Fortunately, White was an untiring correspondent, and many of his findings were recorded in his letters to a handful of friends and acquaintances. One of his most distin-

---

[20]   del Hoyo, J., Elliot, A. & Sargatal, J. eds. (1999). Handbook of the Birds of the World. Vol. 5. Barn-Owls to Hummingbirds. Lynx Edicions, Barcelona. Used with permission.

guished correspondents, the Hon. Daines Barrington, persuaded him to publish some of the letters. This he did, resulting in *The Natural History and Antiquities of Selborne*, appearing in 1789. It was the testament of a modest man, content to deepen his understanding of one small corner of the world, in which he was placed. Even to this day, it remains a treasured text for naturalists and a classic work of English literature.

Common Cuckoo imm. (*Cuculus canorus*)
(*Photo: Pete Walkden*) [21]

2.5 Daines Barrington

As mentioned above, one of White's correspondents was a well-known judge, historian, journalist, and ornithologist by the name of Daines Barrington (1727–1800). He was especially keen to discover how young birds acquired the song of their particular species. He knew that bird keepers taught Bullfinches (*Pyrrhula pyrrhula*), to imitate folk tunes played to them on the flageolet, and so he placed Linnet (*Carduelis cannabina*) chicks into the nests of different birds like Skylarks (*Alauda arvensis*), Wood Larks (*Lullula arborea*), and Meadow Pipits (*Anthus pratensis*). He observed that the fledglings seem to sing rather like their particular foster parents, and from this he concluded that birds picked up songs

---

[21] This image is a reproduction of the photograph taken by Pete Walkden and used with permission. Originally placed on www.flickr.com, the original photograph can be seen at: http://www.flickr.com/photos/petewalkden/4839281688/.

by listening to their "parents", whoever they may be. This proved to be a sweeping generalization, as a Cuckoo *(Cuculus canorus),* did not pick up the song of the host by which it was raised.

> "Everyone wants to understand painting.
> Why is there no attempt to understand the song of the birds?"
>
> **Pablo Picasso (1881–1973)**
> *(Spanish Artist and Painter)*

He realized that he needed to be able to transcribe the songs of birds to be able to analyze them accurately. By that time composers had already incorporated bird songs into their music, albeit with limited success, as the intervals between notes could not be measured and therefore be agreed on. Yet, he contemplated the idea of using musical notation, and even asked a harpsichord tuner to assist him in transcribing what he heard into patterns of crotchets and quavers. He was however doomed to failure, as much of a bird's song defies the human ear, and does not conform to our diatonic (Greek: meaning "progressing through tones") scale. Ornithologists would have to wait until the middle of the twentieth century, when the recording technology had developed to such an extent that useful recordings could be made.

2.6 Lazzaro Spallanzani

Still in the eighteenth century, it was thought that Bats *(Chiroptera),* employed some supernatural power to navigate in the dark. As Professor of Natural Sciences at Pravia University in Italy from 1769 to 1799, and one of the last contributors to the Italian Renaissance, Spallanzani (1729–1799) suspected that they possessed a mysterious sixth sense, and he subsequently issued a challenge to anyone who could discover its nature. The response came from Geneva, where Ludwig Jurine discovered that bats with blocked ears could not detect obstacles in their flight paths. Spallanzani collected bats from the tower of the Pravia cathedral, blinded them all, blocked the ears of some, and discovered that bats needed to hear, and not see, in order to navigate in the dark. It was, however, only in the twentieth century that the ultra-sonic nature of the bat's sound navigational system was discovered.

Charles Waterton (1782–1865)
(Photo: Wikimedia Commons) [22]

2.7 Charles Waterton

After many pioneering journeys to South America, and managing his family's sugar plantations in Guyana in the Caribbean for twenty years, Charles Waterton (1782–1865) inherited the family estate in Yorkshire, and became the twenty-seventh Lord of Walton Hall. It is here that he completed a five kilometer long wall around his estate, to protect the members of "the feathered tribe" against predators and poachers, and so establishing what could be called "the first bird sanctuary in the British Isles."

In it he erected towers for roosting Starlings *(Sturnus vulgaris)*, and build "homes" (the first nesting boxes?) for owls. He converted many hollow trees into observation posts, becoming the first ornithologist to employ a hide in studying bird behavior. He employed advanced techniques for his time. He dissected regurgitated owl pellets in order to determine their diet, (See par. 16.1), and counted the number of visits a pair of Blue Tits *(Parus caeruleus)*, made to their nest in a day while raising their chicks.

Waterton was an early opponent of pollution. He fought a long-running court case against the owners of a soapworks which had been set up near his estate in 1839, and sent out poisonous chemicals which severely damaged the trees in the park and polluted the lake. He was eventually successful in having the soapworks moved.

---

[22] This image is a file from the Wikimedia Commons. Full details of the file can be found at: http://en.wikipedia.org/wiki/File:Charles_Waterton.png. This image (or other media file) is in the public domain because its copyright has expired. This applies to Australia, the European Union and those countries with a copyright term of life of the author plus 70 years.

CHAPTER 3
THE MODERN ERA

> "The path no bird of prey knows,
> nor has the falcon's eye caught sight of it."
>
> **Job 28:7 (NASB)**[1]

Not being a historian, and purely for the purposes of this book, I shall regard the European Industrial Revolution at the end of the eighteenth, beginning of the nineteenth century, as the advent of the Modern Era. Wikipedia lists ninety-one ornithological significant events in the nineteenth century in their article "Timeline of Ornithology".[2]

3.1 John James Audubon

John James Audubon (*Jean-Jacques Audubon*) (1785–1851), "was a French - American ornithologist, naturalist, and painter. He was and is notable for his expansive studies to document all types of American birds and for his detailed illustrations that depicted the birds in their natural habitats. His major work, *The Birds of America* (1827–1839), is considered one of the finest ornithological works. Audubon identified twenty-five new species and a number of new sub-species."[3]

After an extremely eventful live up to then, Audubon embarked on a trip to Mississippi, Alabama, and Florida in search of ornithological specimens on October 12, 1820. His aim was to find and paint all the birds of North America for eventual publication.

---

[1] Scripture quotations taken from the New American Standard Bible®, Copyright © 1960, 1962, 1963, 1968, 1971, 1972, 1973, 1975, 1977, 1995 by The Lockman Foundation. Used by permission. (www.Lockman.org).

[2] Wikipedia contributors. "Timeline of ornithology." Wikipedia, The Free Encyclopedia. Wikipedia, The Free Encyclopedia, 7 Jun. 2011. Web. 31 Aug. 2011.

[3] Wikipedia contributors. "John James Audubon." Wikipedia, The Free Encyclopedia. Wikipedia, The Free Encyclopedia, 30 Jul. 2011. Web. 6 Aug. 2011.

John James Audubon (1785–1851)
(Photo: Wikimedia Commons) [4]

Time and space does not allow us to relate this remarkable story completely here. The result was the publishing of his *Birds of America*. It was first published as a series of sections between 1827 and 1839, in Edinburgh and London. A monumental work consisting of four hundred and thirty-five hand-colored, life-size prints of four hundred and ninety-seven bird species, made from engraved copper plates of various sizes depending on the size of the image. They were printed on sheets called the "double-elephant" paper size and measuring about ninety-nine by sixty-six centimeters. The work contains just over seven hundred North American bird species. The first and perhaps most famous plate was the Wild Turkey (*Meleagris gallopavo*), which had been Benjamin Franklin's candidate for the national bird. It lost to the Bald Eagle (*Haliaeetus leucocephalus*).

"I wish the bald eagle had not been chosen as the representative of our country; he is a bird of bad moral character . . . like those among men who live by sharking and robbing, he is generally poor, and often very lousy . . . the turkey is a much more respectable bird, and withal a true original native of America."

Benjamin Franklin (1706–1790)
(American Statesman, Scientist, Philosopher, Printer, Writer, and Inventor)

Audubon's aim was to paint the birds in life size. For many birds even the so-called "double-elephant" paper size was too small. It is the way in which Audubon solved this dilemma that eventually became his hallmark. He simply portrayed the larger birds in

---

[4] This image is a file from the Wikimedia Commons. Full details of the file can be found at: http://en.wikipedia.org/wiki/File:John_James_Audubon_1826.jpg. This image (or other media file) is in the public domain because its copyright has expired. This applies to Australia, the European Union and those countries with a copyright term of life of the author plus 70 years.

positions which would fit the book's format. The results were sometimes bizarre and even perplexing images. For example: the Great Blue Heron (*Ardea herodias*). The legs are slightly bent, and the neck is curved in a flat arc downwards, so that the tip of the beak points into the lower corner of the picture. His work includes images of six birds that are now-extinct: the Carolina Parakeet (*Conuropsis carolinensis*), Passenger Pigeon (*Ectopistes migratorius*), Labrador Duck *(Camptorhynchus labradorius),* Great Auk *(Pinguinus impennis),* Esquimaux Curlew (Eskimo Curlew/Northern Curlew) *(Numenius borealis),* and Pinnated Grouse (Prairie Chicken) *(Tympanuchus Spp.).* Audubon's work was, to say the least, a remarkable accomplishment. It took more than fourteen years of field observations and drawings, plus his single-handed management and promotion of the project to make it a success.

Audubon sold oil-painted copies of the drawings to make extra money and publicize the book. He also had his portrait painted by John Syme, who clothed the naturalist in frontier clothes promoting his rustic image. The portrait was hung at the entrance of his exhibitions and now hangs in the White House.

King George IV was also an avid fan of Audubon and a subscriber to the book. Audubon was elected to the Royal Society of Edinburgh, the Linnaean Society, and London's Royal Society in recognition of his contributions. At the latter he followed Benjamin Franklin, who was the first American fellow. Audubon was elected a Fellow of the American Academy of Arts and Sciences in 1830. In 1905, the National Audubon Society was incorporated and named in his honor. Its mission is "to conserve and restore natural ecosystems, focusing on birds..." On 6 December, 2010, a copy of *Birds of America* was sold at a Sotheby's auction for US$11.5 million, a record price for a single printed book.

In 1839 having finished his *Ornithological Biography*, Audubon returned to the United States with his family. He bought an estate on the Hudson River (now Audubon Park). In 1842, he published an octavo edition (one eighth the size of the original sheets) of *Birds of America*, with sixty-five additional plates. It earned US$36,000 and was purchased by one thousand one hundred subscribers. Audubon spent much time on "subscription gathering trips", drumming up sales of the octavo edition, as he hoped to leave his family a sizable income. He died at his family home on January 27, 1851.

While living and working in the city of Shenyang in the north of China, early in 2003, I was scrounging around second hand book stores, when, to my delight, I came upon a copy of a book called *John James Audubon, Birds of America*. The text was written by a Helgard Reichholf-Riehm. It was published by Benedikt Taschen Verlag GmbH in 1994 in Germany. It contains ninety-five pages filled with photographs to illustrate Audubon's style and technique. On page two is a reproduction of a portrait of Audubon, courtesy of The White House Collection, Washington D.C.

While in Edinburgh to seek subscriptions for the book, Audubon visited Professor Robert Jameson's Wernerian Natural History Association. He gave a demonstration of his method of propping up birds with wire when painting them. In the audience was a young student called Charles Darwin.

3.2 Charles Darwin

The "mystery of mysteries" at this point in time in the scientific world was the origin of different species, including that of the human species. Charles Robert Darwin (1809–1882), was the first of a new generation of scientists and naturalists who had an insatiable desire to understand and objectively explain what they observed in their world. Darwin was one of them, and an untiring and methodical researcher who was of the opinion that field observations alone were useless, unless one could connect them to a theory or hypothesis of some kind.

On the 27th of December, 1831, Darwin embarked on a five-year long cruise on the British scientific vessel, the *H.M.S. Beagle* along the coastline of South America. *The Voyage of the Beagle* is a title commonly given to his first book that was published in 1839 as his *Journal and Remarks*, bringing him considerable fame and respect. In the Galapagos he noted that every island had its own variety of birds, and in Patagonia he made numerous fossil discoveries.

Charles Robert Darwin (1809–1882)
(Photo: Wikimedia Commons) [5]

In Chile he commented on the aptness of the name Tapacolo (also spelled Tapaculo) given to the family *Rhinocryptidae* by saying: "It is called Tapacolo, or 'cover your posterior'; and well does the shameless little bird deserve its name; for it carries its tail more than erect, that is, inclined backward towards its head." He became more and

---

[5]   This image is a file from the Wikimedia Commons. Full details of the file can be found at: http://upload.wikimedia.org/wikipedia/commons/0/0f/Charles_Darwin_by_Maull_and_Polyblank%2C_1855-1.jpg. This image (or other media file) is in the public domain because its copyright has expired. This applies to Australia, the European Union and those countries with a copyright term of life of the author plus 70 years.

more interested in the problem of the changes that took place within a species. He pondered over the work of The Reverend Thomas R. Malthus (1766–1834) entitled *On Population,* in which Malthus concludes that "the fittest survive." Although a number of scientists before Darwin suggested the possibility of an evolutionary process, he perceived the so-called "descent with modification" when he realized that the environment in which a creature lives ultimately promotes continuous development in its traits. This he found is indeed true of all living creatures. According to Darwin, these minute differences in individuals are perpetuated and even accentu-

Greylag geese *(Anser anser)*
(Photo: Ted van den Bergh) [6]

ated and increased in the descendants of that particular creature. To Darwin, the fact that animals, as well as plants, produce more offspring than could possibly survive implied that there has to be a "struggle for existence." He was the first to provide a plausible theoretical framework for the forces that keep on driving this process forward. He called it "natural selection." As he assumed that humans were part of this process, he knew the implication was that a random process replaced God as the Creator of species and of man. Because of this, he was initially reluctant to publish his theory.

On 18 June 1858 he received a letter from a naturalist Alfred Wallace, accompanied by an essay by Wallace entitled: *Essay on the Tendency of Varieties to depart indefinitely from the Original Type.* His first reaction was to publish Wallace's essay, and accept failure. But for the insistence of his friends, and by one of the greatest acts of generosity known among scientists, Wallace withdrew his theory and allowed Darwin to present his essay entitled: *The Origin of Species by means of Natural Selection, or the Preservation of favored Races in the Struggle for Life,* which became shortened to *On the Origin of Species,* to the Linnean Society on 1 July 1858. It was actually presented by two friends:

---

[6]   This image is a reproduction of the photograph taken by Ted van den Bergh and used with permission. Originally placed on www.flickr.com, under the username "webted", the original photograph can be viewed at: http://www.flickr.com/photos/webted/5712895344/in/set-72157605244874930.

Sir Charles Lyell and Sir Joseph Dalton Hooker, as Darwin's baby son died from scarlet fever at that time and Darwin was too distraught to attend personally. It was soon realized that it was a revolutionary idea of the first magnitude. The one thousand two hundred and fifty copies of the first edition of the book were sold out on the day of its publication, on 22 November 1859. Within a few years it had been translated into almost every civilized language.

The apparent attack on the validity of the Scriptures predictably caused uproar in the Church. Darwin was immediately attacked from all sides. The mistaken view was that Darwin was saying that man descended from another primate, and this caused the outrage. However, great scientists like Thomas Henry Huxley and Charles Lyell supported him. Huxley, a powerful writer and a brilliant debater, was Darwin's chief supporter. The defeat by Huxley of the over-confident Bishop of Oxford in the British Association debate on 30 June 1860 marked a watershed in the acceptance of the theory. In 1871 Darwin published his *Descent of Man,* a further elaboration of the evolution theory, attempting to correct the mistaken views of his critics. In 1872 he published *The Expressions of Emotions in Man and Animals,* the first major scientific work devoted to behavior. It established the foundations for the two separate disciplines of experimental psychology and ethology. (See par. 3.5). His other principle works were *Insectivorous Plants* (1875), *Different Forms of Flowers* (1877), and *Worms* (1881).

Charles Darwin died on 19 April 1882, and his wish was to be buried in the churchyard of the Kentish village of Down, were he had lived the last forty years of his life. But, at last the authorities and the people of his home country made a gesture of recognition to his greatness, and he was buried in Westminster Abbey, just a few feet from that other great scientific discoverer, Isaac Newton.

3.2.1 A Personal Note.

Sometimes being a layman has its advantages. I have no problem reconciling Darwin and my Bibles. I cannot explain how the universe and its creatures were created. I believe my Bibles when they state that a "Creator" was involved, and that "He" created it from "nothing" in six "days." I have to admit that there are an enormous amount of issues that I simply cannot get to grips with. I don't even know exactly what my Bibles means by "nothing." In addition, I have no idea what my Bibles mean when they talk about a "day." I believe science has proved that our universe and species have developed over millennia. I do believe that one, and only one, of those species, *Homo sapiens,* meaning "wise man" or "knowing man" developed into something unlike any other species.

"Humans have a highly developed brain, capable of abstract reasoning, language, introspection, and problem solving. This mental capability, combined with an erect body carriage that frees the hands for manipulating objects, has allowed humans to make far greater use of tools than any other living species on Earth. Other higher-level thought processes of humans, such as self-awareness, rationality, and sapience, are considered to be defining features of what constitutes a 'person'."[7]

I do, however, believe we should add one more distinctive feature when defining a "person." This is coupled to my view on the concept of "eternity." Staring up into the night-sky, and taking note of what science has to date established as facts, I cannot but believe that space is eternal. There is no end. One can travel for ever but will not be able to get to a fence with a sign attached to it saying: "The End." Even if there were such a point, we could question what lies beyond that point. So, if we can travel for ever, it means "for ever" in terms of time as well. To me that means "eternity." Then, in my mind, the feature that distinguishes *Homo sapiens* definitively from other species is that we have a "hope" of experiencing eternity in some way, state, or form.

How did we receive this "hope"? I believe that at some point during the millennia of evolution, humans received an addition trait over and above that of a physical body. This was not the case in any other species, and contributes to our uniqueness. We call it a "soul", but what exactly that means no one knows. I do however believe that this mysterious entity gives us the capability to exist "for ever" in some way, state, or form. In this regard, the discovery of the Neanderthal man (an extinct member of the *Homo* genus) is an interesting development in our knowledge of the past, and could possible in future assist in our understanding of whom exactly we are.

How, what, and where this "eternal life" will be, I believe no one knows. I can appreciate the quest of those individuals who spend enormous amounts of time, money, and effort to find some answers, but I believe it is a futile exercise. We can speculate, guess, fantasize, and dream, as much as we like, but unfortunately we will not be any the wiser from within our present reality. The simple fact is: we cannot go there in terms of space, time, or any other dimension, and then return to our situation and report back.

Why do we have this "hope"? According to my Bibles we were made so special so that we would be capable of spending eternity in the presence of the Originator, applauding him/her for giving us this unique opportunity called "life." Why do only some of us believe this? I simply don't know. As I said before: "Being a layman has its advantages."

Faith in God & his promises!

---

7   Wikipedia contributors. "Human." Wikipedia, The Free Encyclopedia. Wikipedia, The Free Encyclopedia, 5 Aug. 2011. Web. 6 Aug. 2011.

> My heart is not proud, LORD, my eyes are not haughty;
> I do not concern myself with great matters or things too wonderful for me.
> **Psalm 131:1 (NIV)[8]**

3.3 Edward Lear

Edward Lear was an English artist, illustrator, author, humorist, and poet. In the 1830's he became one of the world's greatest bird artists. He illustrated *The Family of the Psittacidae* (Parrots) for the London Zoological Society in 1832 and also John Gould's *The Birds of Europe* (1832–1837). His paintings are often favorably compared with those of Audubon. He was especially fascinated by owls, and although he quitted ornithological illustration in 1837, he never forgot his beloved owls. This is evident from his *The Owl and the Pussycat* which he wrote in 1868 and was published in 1870. His depiction of the Eurasian Eagle-owl *(Bubo bubo)* in these publications is "today the most famous and sought-after plate among the 3000 illustrations in Gould's works" (del Hoyo, et al, Vol. 5, p140).[9] He traveled throughout Europe and the Near East from 1837 to 1847, recording his travels in his *Illustrated Journals of a Landscape Painter*. Later he published *A Book of Nonsense* (1846), *Nonsense Songs* (1871), and *Laughable Lyrics* (1877), establishing himself as the master of limerick, to which he gave the modern formula and metric cadence.

3.4 The Behaviourists

Behaviorism, noun: (Psychology), the study of human actions by analyzing stimulus and response; the doctrine stating that such study is the only valid method in psychology.

Douglas Alexander Spalding (1841–1877), working from Ravenscroft, the home of the Amberley family in the Welsh border country, discovered that flying was a skill that was gained as the bird's nervous system developed. By experiments with domestic chicks and piglets, he correctly identified "imprinting", and was convinced that behavior was in a sense "automatic", produced by inborn nervous processes.

---

[8] THE HOLY BIBLE, NEW INTERNATIONAL VERSION®, NIV® Copyright © 1973, 1978, 1984, 2011 by Biblica, Inc.™ Used by permission. All rights reserved worldwide.

[9] del Hoyo, J., Elliot, A. & Sargatal, J. eds. (1999). Handbook of the Birds of the World. Vol. 5. Barn-Owls to Hummingbirds. Lynx Edicions, Barcelona. Used with permission.

The German Jacques Loeb (1859–1924) eradicated the last traces of conscious mind from interpretations of animal behavior. Loeb learned that Julius von Sachs spoke about non-voluntary actions by plants as *tropisms*. Loeb in turn confirmed that this particular phenomenon occurred in the animal world as well.

The Russian physiologist, Ivan Petrovich Pavlov (1849–1936) received the Nobel Prize in 1904, at the age of fifty-five. He discovered conditioned reflexes by observing the digestive systems of Russian hounds. The American Edward Lee Thorndike rewarded "correct" behavior by his animals, and formulated his *law of effect* which stated that "behavior changes because of its consequences."

It was, however, another American, John B. Watson (1878–1958), who de-scribed a theoretical framework of ex-

Short-eared Owl (*Asio flammeus*)
(*Photo: Pete Walkden*) [10]

perimental psychology, called *Behaviorism*. On Bird Key, one of the Tortuga Islands off Florida, he found that Sooty Terns (*Sterna fuscata*) rigidly adhered to a habit, ignoring changes in their surroundings, to find their nests in a colony. He moved a tern's egg about thirty centimeters, to a new spot, but on returning, the bird would continue to sit on the now empty nest, in full view of its egg.

B. F. Skinner (1904–1990) introduced secondary reinforcement as a technique for manipulating behavior and promoting learning. He was successful in implementing his principles into the advertising world. Some of his techniques, like flashing millisecond ads for Coca-Cola, were however banned from being used in cinemas. But zoologists still believed that behavior was shaped by evolution and not by regimes of rewards. Konrad Lorenz, for example, observed Greylag Geese *(Anser anser)* on the banks of the Danube,

---

[10] This image is a reproduction of the photograph taken by Pete Walkden and used with permission. Originally placed on www.flickr.com, the original photograph can be viewed at: http://www.flickr.com/photos/petewalkden/4013937542/in/set-72157622834937469/.

and found that the bond between a gander and her goslings were forged independently of rewards.

By the 1950's portable tape recorders were introduced, and the Bell Telephone Company had invented the sound spectrograph. It is able to do what the prism does for light. It digests a noise, sorts out the range of frequencies, and projects them in printed form. William Thorpe (1902–1986), from the Madingley Ornithological Field Station just outside Cambridge at the time, borrowed the only one in the UK from the Admiralty, and found that Chaffinches *(Fringilla coelebs)* were born with a rough vocal prototype, a "genetic blueprint", which was developed by listening to their parents.

Over the past two decades, the entrenched views of experimental psychologists have mellowed to some extent under the influence of a generation of more moderate and pragmatic scientists. They are realizing that behavior is the result of a blend between nature and nurture. In other words, most habits are learned within constrains imposed by an animal's genes.

3.5 Ethology

Ethology, noun: the science of character-formation; the science of studying animal behavior.

By the start of the twentieth century, accurate accounts of bird behavior hardly existed. Julian Huxley, the grandson of Darwin's bulldog, in 1912, studied Great Crested Grebes *(Podiceps cristatus),* and came up with the term *ritualisation.* He realized that displays were fully or in part composed of common everyday actions, refashioned by natural selection to enhance their effectiveness as signals employed in communication. Oskar Heinroth, Director of the Berlin Zoo's aquarium, in 1933, published his *The Birds of Central Europe,* finally establishing that each kind of bird has its own unique set of behavioral patterns. He laid the foundation for a science of comparative behavior, which mirrored and complemented the discipline of comparative anatomy, and called it *ethology*.

Konrad Zacharias Lorenz (1903–1989) "was a Nobel Prize winning Austrian zoologist, ethologist, and ornithologist. He shared the 1973 Nobel Prize with Nikolaas Tinbergen and Karl von Frisch. He is often regarded as one of the founders of modern ethology, developing an approach that began with an earlier generation, including his teacher Oskar Heinroth. Lorenz studied instinctive behavior in animals, especially in Greylag Geese *(Anser anser),* and Jackdaws *(Coloeus monedula).* Working with geese, he rediscovered the principle of imprinting (originally described by Douglas Spalding in the

19th century) in the behavior of nidifugous (those that leave the nest shortly after hatching or birth) birds."[11]

From 1935 to 1938 Konrad Lorenz, studied Greylag Geese *(Anser anser)*, Eiders *(Somateria mollissima)*, Teal *(Anas crecca)*, and Goldeneyes *(Bucephala clangula)* (a seaduck), on the ponds around Altenberg, east of Vienna. He found that the drakes of different species each had its own way of displaying to the ducks. Together with the young Dutch ornithologist, Nikolaas Tinbergen, they found that the young goslings reacted only to selected features of what they see. These are called "releasers". This was confirmation of what falconers like Frederick II knew and utilized seven hundred years and more ago. Another example is the pronounced red spot on the lower mandible on the otherwise yellow beak of the parent Herring Gull *(Larus argentatus)*. The red spot is the "releaser" that prompts the chick to beg for food from the parent. Tinbergen found that by exaggerating the contrasting colors some species form a "supernormal releaser", which for example is used effectively by the Cuckoo *(Cuculidae)* chick in prompting his foster parents to feed him first. They do this by flashing a relatively huge and irresistibly brightly colored gape (the interior of the open mouth) to the foster parent.

> "There'll be bluebirds over
> The white cliffs of Dover,
> Tomorrow, just you wait and see."
>
> **Nat Burton, born Nat Schwartz (1901–1945)**
> *(American songwriter)*
> From the song: *The White Cliffs of Dover.*

After the Second World War ornithologists began to focus their attention on what actually caused behavior, its function(s), and how it evolved. Stress, in a negative sense (e.g. protecting your territory from an intruder), as well as in a positive sense (e.g. in courting), was a major origin of arresting displays. Behavioral deadlocks do occur when, for example, the above intruder is an amorous female, placing the male in an emotional turmoil. This situation often leads to irrelevant behavior before copulating. For example: the preening of the Avocet *(Recurvirostra avosetta)*, the tit-bitting of the Red Junglefowl *(Gallus gallus)*, and the bow or curtsey of the Indian Peafowl *(Pavo cristatus)*.

In 1952, the quest by ornithologists to determine the "function of behavior", or "signals for survival", led to the appearance of the first woman to become prominent in the

---

[11] Wikipedia contributors. "Konrad Lorenz." Wikipedia, The Free Encyclopedia. Wikipedia, The Free Encyclopedia, 23 Jul. 2011. Web. 6 Aug. 2011.

ornithological world. Three years earlier Tinbergen had moved to Oxford University, and now he acquired as a student, a post-graduate from Switzerland by the name of Esther Cullen. He persuaded her to study the Kittiwakes *(Rissa tridactyla)* of the Fame Islands, three centuries after John Ray and Francis Willughby. (See par. 2.2). She found that although Kittiwakes were extraordinarily quarrelsome when compared to Herring Gulls *(Larus Spp.)*, they were completely tame, some even allowing her to touch them. They had completely lost the need to take flight as their nests were plastered onto cramped ledges, making them virtually invulnerable to predators. However, because spaces on a suitable rocky shelve was at a premium, a high level of territorial hostility occurred between them. In their case, courtship feeding is directly into the gullet of the female, preventing it from falling into the sea. Also, the female would be sitting down rather than standing up during copulation. These and many other discoveries by Cullen showed how effectively their behavior had been adapted by evolution's natural selection to life on the dangerous ledges. She published her findings in *Ibis* (The International Journal of Avian Science), Volume 99, April 1958, under the title *Adaptations in the Kittiwake to cliff-nesting*. Other ornithologists confirmed this by for example observing that birds living in dark thickets would resort to audio rather than visual signaling, often evolving into drab brown birds, but highly melodious songsters. For example: the Babblers *(Timaliidae)*, and the Nightingale *(Luscinia megarhynchos)*. Visual signaling, on the contrary, becomes more important when attention has to be drawn over the long distances encountered in the open country like plaines, grasslands, and savanna. For example: in the case of the Ostrich *(Struthio camelus)*, the Cranes *(Gruidae)*, and the Bustards *(Otididae)*.

3.6 Animal societies

Travel to the Bushveld of Southern Africa, and one can observe an awe inspiring ornithological sight. Millions of Redbilled Quelea *(Quelea quelea)*, in a densely packed flock, and in highly synchronized flight over the African savanna, by times virtually blocking out the fierce African sun. Or travel to the Atlantic coast in that part of the continent, and see tens of thousands of Gannets *(Sulidae)*, nesting at beak's length, with the stifling smell of their guano, and the deafening noise of their chatter. These are but two examples of some of the most wonderful spectacles in nature because "birds of a feather flock together."

Ornithologists and zoologists are faced with an array of different kinds of animal communities, the insects being the masters. Capable of acts of apparent heroism, individuals will in instinctive cooperative behavior and completely unselfishly, literally give their lives in the service of their colony. In this vein, A.V. Espinas, in 1878, concluded that no living being was truly solitary, with even the most steadfast "loners" forming temporary attachments from time to time.

Eugene Nielsen Marais (1871–1936), was South Africa's greatest but also most tormented polymath. He was a naturalist to rival the best in the world, a self-taught surgeon, prospector, lawyer, journalist, writer and poet. It was his observations and writings on termitaries, birds, baboons, and the human psyche that most set him apart. He became a newspaper editor at the age of nineteen, and studied law in London from 1896 to 1902. Back in South-Africa he wrote *My Friends the Baboons*, *The Soul of the White Ant* and *The Soul of the Ape*, suggesting in the former the idea that a termitary (nest of termites/white ants) might form a unity similar to a highly developed animal, with the queen literary being the brain. The only difference be-

Eugene Nielsen Marais (1871–1936)
(Photo: Wikimedia Commons) [12]

tween the termitary and other organisms was that the former could not move. In addition he wrote about his observations on birds. He was a pioneer in the field of zoology by being the first to observe intensively a group of primates in their natural surroundings.

He had read Darwin, and in an unpublished article, *The Yellow Streak in South Africa*, he seems to agree to some extent with those who have wondered whether the hunter-gatherer peoples of southern Africa called the Bushman or San people can really be called human beings. In 1982, Leon Rousseau published a biography on Marais called *The Dark Stream—The Story of Eugene Marais*. In it he says that Marais' disposition towards other races like the Bushman can be described as antipodal to the anthropological snobbery and arrogance which made the Victorians react to Darwin's theories in such a highly emotional way. (*The Dark Stream*, p. 267).

By hypnotizing willing friends, Marais found that humans still possessed their ancient sensory abilities, albeit latent. When a Czech, Rezl, managed to accentuate extrasensory perception by means of hypnosis in 1966, it was regarded as a breakthrough in

---

[12] This image is a file from the Wikimedia Commons. Full details of the file can be found at: http://en.wikipedia.org/wiki/File:Eugenemarais.jpg. This image (or other media file) is in the public domain because its copyright has expired. This applies to Australia, the European Union and those countries with a copyright term of life of the author plus 70 years.

the world of parapsychology. Sixty years earlier Marais found that under hypnosis a girls hearing ability became eight times more acute, surpassing that of a baboon. To prove the persistence of the phyletic (evolutionary descent) memory in birds, he bred four generations of Weavers *(Ploceus Spp.)* by hatching them under Canaries *(Fringillidae)*. He committed suicide and so ultimately his life was a tragic tale of genius. His manuscripts were lost for about fifty years, and were eventually published in the late 1960's.

A Norwegian, Thorleif Schjelderup-Ebbe (1894–1982?), measured the social relationships amongst domestic chickens, and observed an avian ranking system or "pecking-order." The assertive behavior has obvious survival value to those at the top, as they have the first choice of food and preferential access to mates. He discovered *despotism* as a force to maintain order in communities. He applied a technique just developed by Scandinavian naturalists, and marked individuals by placing colored rings or bands on their legs. By quantifying behavior, he showed that science and ornithology thrives on statistics.

But there are communities in nature that order themselves differently. Most seabirds and the majority of hoofed mammals re-order themselves for breeding. Frank Fraser Darling (1903–1979) was one of the first to describe such structural alterations. He studied the Red Deer *(Cervus elaphus)* of Scotland, seabirds, and Grey Seals *(Halichoerus grypus)*. Prey species often tend towards some kind of gregariousness, for example, in the case of small birds like Finches *(Ploceidae)* who close ranks when sighting a bird of prey. A host of birds feed better in flocks. Starlings *(Sturnidae)* for example act as beaters for each other. Cormorants *(Phalacrocoracidae)* drive their prey before them when fishing in flocks. Colonial nesting of sea birds act as erotic theaters, where the huge numbers of displaying individuals raise the level of excitement resulting in synchronous breeding and a sudden glut of eggs and chicks, impossible to devour by predators.

Israeli ornithologist, Amotz Zahavi was one of the founders of the *Society for the Protection of Nature in Israel*. As he studied a clan of about twenty Arabian Babblers *(Turdoides squamiceps),* he found that the family structure could be much more intricate. The Babblers were basically an extended family each containing, on average, a couple of mature males, a pair of females who produced a joint clutch, a number of yearlings and fledglings. Each of these groups had definite functions, e.g. the non-breeding yearlings assisted their elders in finding and supplying the brood with food, but with all partaking in some activities e.g. defending their territory.

We have now covered the development of ornithology over some forty centuries, bringing us to the twentieth and twenty-first centuries AD. In the next Chapter we shall take note of the latest developments.

CHAPTER 4
MODERN DEVELOPMENTS

> "Is my inheritance like a speckled bird of prey to me?
> Are the birds of prey against her on every side?"
>
> **Jer 12:9 (NASB)**[1]

4.1 Introduction

Within the scope of this book it would be impossible to fully discuss all aspects of all the developments taking place in the field of Ornithology today. In its article "Timeline of ornithology",[2] Wikipedia lists no less than one hundred and twelve significant ornithological dates in the twentieth century. I have, therefore, selected a few which would surely be of interest to relatively new birders, and am hoping that it would prompt them to read further, and even to become actively involved in one or more of those fields that interest them most.

Ornithologists are employing the latest technological developments to assist them in reaching conclusions sooner and more cost effectively, and which, in addition, are more reliable than ever before. The developments in computer technology in general, and more specifically, the Internet, would probably be one of the most obvious. For example, radio transmitters, web cams, and satellite-tracking are some of the latest tools in studying the movements of birds on the Net.

To confirm the above, one only has to look at the development of bird publications. Field guides have become more and more functional, with the standard of the texts, the artworks, their reproduction into illustrations, and the general printing techniques continually improving. I am sure that one could visit any place in the world and

---

[1] Scripture quotations taken from the New American Standard Bible®, Copyright © 1960, 1962, 1963, 1968, 1971, 1972, 1973, 1975, 1977, 1995 by The Lockman Foundation. Used by permission. (www.Lockman.org).

[2] Wikipedia contributors. "Timeline of ornithology." Wikipedia, The Free Encyclopedia. Wikipedia, The Free Encyclopedia, 7 Jun. 2011. Web. 31 Aug. 2011.

find at least one, if not more, functional field guides covering the country involved, or even a specific area.

I can speak from personal experience. As soon as I arrived in the city of Shenyang in the north of China in October 2002, I visited a large bookstore (purportedly the largest in South-East Asia at the time). After finding the English section I started explaining the concept of a field guide to the staff. And, lo and behold, there was a book entitled in Chinese with an English subtitle *Atlas of Birds of China*, published in 1995. As I inspected it I found that it was not new (which didn't bother me), but to my dismay, that pages 65–80 were missing. The text was in Chinese, but it contained clear illustrations so I decided to buy it even though I was told it was for sale at the original price only. Five years later, while browsing through a very small, very smelly, very dusty, second-hand bookshop in the city of Shenzhen, I found another copy of the same book, with all the pages intact, and promptly bought it (for next to nothing, I might add).

The recently published volumes of *The Handbook of the Birds of the World* must be the ultimate in teamwork, dedication and technical excellence. I for one have been saved many hours of library work by the comprehensiveness of this mammoth task. On their website, www.lynxeds.com, the publishers, Lynx Edicions in my view correctly state that the *Handbook of the Birds of the World* is the first work ever to illustrate and deal in detail with all the living species of birds. It goes on to say that when Volume 16 is published in 2011, (Volume 1 was published in 1992), it will be the first work to verbally and visually portray each member of an entire Class of the Animal Kingdom.

I do not know of any other science, which, on a worldwide scale, utilizes the time, effort, funds, and expertise of amateurs to the same degree as ornithology. The extent of studies undertaken by amateurs working within the parameters of strict scientific methodology is in fact unique to ornithology. Their integrity in assisting the professionals by collecting data *in loco* for example, has long ago been placed above any doubt. One of the reasons for this achievement is the fact that birders vehemently adhere to birdwatching etiquette like for example the universal rule: "If in doubt, don't."

Atlas projects, bird ringing, bird rehabilitation, reserve development, etc., are a living testimony to those who have put their money, time, and efforts were their mouth is. For example: The South African Bird Atlas Project (SABAP1), covered six countries, and 3103 topographic grid cells; more than five thousand active volunteer observers were involved in intensive data collection over a period of six years; nine hundred and thirty species were involved; 149 434 checklists containing 7 332 504 records and 243 248 breeding records submitted; a covering of ninety-seven percent of the cells were achieved. SABAP2 is the acronym for the second Southern African Bird Atlas Project and is the follow-up on SABAP1. The first atlas project took place from 1987–1991. The current project wants to map the distribution and abundance of birds in Southern Africa and the atlas area includes South Africa, Lesotho and Swaziland. The second atlas project started on 1 July 2007 and will run to 2011.

> "Bird watching is a bloodless expression of man's primitive hunting instincts. We have substituted binoculars and cameras for the gun, but we still seek a trophy: A new species on a life list, or photographs of one of earth's rarest and most exquisite creatures. Our search may take us no farther than a nearby meadow or it can lead to the cloud forests of tropical mountains."
>
> **James A. Kern**

Years ago I had the privilege of observing a pair of Taita Falcons (*Felco fasciinucha*) in the wild, at a spot called Strydom's Tunnel in the north-eastern mountains of South Africa. A group of birders had gathered soon after sunrise, set up there telescopes, and were delighted to watch the birds high up on a cliff ledge. Some folk had come from the surrounding towns, villages, and farms. Others like my friends and I, had traveled extensive distances from the city to get a glimpse of this famous (in birding circles) feathered couple. Nothing extraordinary in that, one would think. Except that one of the young men present, while peering intently through a telescope, casually informed us that he was in a bit of a hurry, as he was getting married to a local lass later that day! I would politely refrain from any comments. Refer to the article "Birdwatching" in Wikipedia for a detailed and interesting discussion on the subject, and a better understanding of the very special people called "birders."[3]

4.2 Migration

Migrate: verb, intransitive: move from one place of residence to another at some distance, especially from one country to another; (birds, fish) come and go with the seasons; move under natural forces.

As one of the most impressive manifestations of animal behavior, migration amongst birds are the most striking. This phenomenon has since the earliest days, fascinated ornithologists, zoologists, birders, and laymen alike. It is characterized by its regularity and predictability, and the fact that it, in the case of birds, affects whole populations of a particular species. It is estimated that forty percent of the birds breeding in the Pale-

---

3   Wikipedia contributors. "Birdwatching." Wikipedia, The Free Encyclopedia. Wikipedia, The Free Encyclopedia, 3 Aug. 2011. Web. 6 Aug. 2011.

Arctic Terns *(Sterna paradisaea)*
(Photo: Wikimedia Commons) [4]

arctic region, winter in the Afrotropical Region, totaling in the order of five billion birds, excluding the water birds. Migration is usually latitudinal, covering long distances, but can also be altitudinal, usually over short distances. Many species follow regular routes across continents and oceans to escape the adverse winter conditions. Arctic Terns (*Sterna paradisaea*), for example, cover at least forty thousand kilometers per annum between the Arctic and Antarctic regions.

Birds weighing as little as eight grams, like the Willow Warbler (*Phylloscopus trochilus*), double their weight in a week or two before migrating. This will ensure sufficient energy to sustain themselves for a non-stop flight of up to one hundred hours, while flying ten thousand kilometers or more at supersonic altitudes (from one thousand to five thousand meters high). Add to this the incredible precision with which they repeatedly reach their destinations every year, and it is no wonder that this phenomenon has caught the attention of man since ancient times.

> "How do geese know when to fly to the sun? Who tells them the seasons? How do we, humans know when it is time to move on? As with the migrant birds, so surely with us, there is a voice within if only we would listen to it, that tells us certainly when to go forth into the unknown."
>
> **Elisabeth Kubler-Ross MD (1926–2004)**
> **(Swiss-American Psychiatrist, and author of *On death and Dying* (1969)**

Most migratory Passerines *(Passeriformes),* many Waders (called shorebirds in North America), Rails *(Rallidae),* and Ducks *(Anatidae)* move only by night, alone or in small

---

[4] This image is a file from the Wikimedia Commons. Full details of the file can be found at: http://en.wikipedia.org/wiki/File:Arctic_terns.jpg. This file is in the public domain, because the authors Toivo Toivanen & Tiina Toppila provided the image "for unlimited free use" on the source site. In case this is not legally possible: The right to use this work is granted to anyone for any purpose, without any conditions, unless such conditions are required by law.

flocks, and on a broad front. Conversely, the soarers e.g. Storks *(Ciconhidae)*, Pelicans *(Pelecanidae)*, and Raptors *(Falconiformes)*, who utilize thermals and follow landmarks, and the Swifts *(Apodidae)*, and Swallows *(Hirundinidae)* who feed in flight, all migrate during the day.

Navigational skills of migratory birds are as yet not fully comprehended by us. The sun, stars, and the earth's magnetic field have been found to assist birds in navigating. It is however only in the last few decades that we have been able to discover the details of the migratory movements in a substantial number of species, and begin to understand the basics involved. This has been due to extensive ringing projects, and the utilization of space-age technology, including satellite tracking. Most of this behavior is inherited genetically, the juveniles usually departing later than their parents, expanding their inherent "program" by the experience gained on the first and subsequent trips. Refer to the Wikipedia article "Bird migration"[5] for much more on the subject, dealing for example with the famous *Emlen funnel.*

4.2.1 The Great Migration

> "Even the stork in the sky knows her appointed seasons, and the dove, the swift and the thrush observe the time of their migration. But my people do not know the requirements of the LORD."
>
> **Jer 8:7 (NIV)[6]**

No discussion on migration will be complete without mentioning the Wildebeest *(Connochaetes)*. It is also called "gnu" from the Khoikhoi name for these animals. *Wildebeest* is Dutch and Afrikaans for "wild beast" or "wild cattle." In Afrikaans *beest* means "cattle", while *Connochaetes* derives from the Greek words *konnos* meaning "beard", and *khaite* meaning "flowing hair."

"Wildebeest are known for their annual migration to new pastures. Wildebeest usually begin their migration in the months of May or June when drought forces them to go on the move. However, if it is a particularly dry year, they may begin their migration

---

5   Wikipedia contributors. "Bird migration." Wikipedia, The Free Encyclopedia. Wikipedia, The Free Encyclopedia, 3 Jul. 2011. Web. 6 Aug. 2011.

6   THE HOLY BIBLE, NEW INTERNATIONAL VERSION®, NIV® Copyright © 1973, 1978, 1984, 2011 by Biblica, Inc.™ Used by permission. All rights reserved worldwide.

earlier than usual because of the decrease in vegetation. The reason that the wildebeest is a migratory animal is because the grass that it feeds on is not a very good provider of energy and minerals, so it has to move around in order to get adequate nutrition. Factors that affect migration include food, water, predators, and also phosphorus level."[7]

The Serengeti ecosystem is a vast geographical region in Africa. It is located in north Tanzania and extends to south-western Kenya. It spans some thirty thousand square kilometers. The Serengeti hosts the largest mammal migration in the world, the so-called "Great Migration", which is one of the ten "Natural travel wonders of the world."

Around October, nearly two million herbivores travel from the northern hills toward the southern plains, crossing the Mara River, in pursuit of the rains. These herds comprise mainly of the Blue Wildebeest, or Brindled Gnu (*Connochaetes taurinus*), and the Plains Zebra (*Equus quagga*, formerly *Equus burchelli*). Zebras and wildebeest group together in open savannah environments when there is a high chance of predation. The Black Wildebeest, or White-tailed Gnu, (*Connochaetes gnou*) has a more southerly range in Africa. In April, the herds then return to the north through the west, once again crossing the Mara River. Some two hundred and fifty thousand wildebeest die during the journey from Tanzania to the Maasai Mara Reserve in lower Kenya, a total of eight hundred kilometers. Death is usually from a range of factors including thirst, hunger, exhaustion, or predation.

"Numerous documentaries feature wildebeest crossing rivers, many being eaten by crocodiles or drowning in the attempt. For example, the 1994 documentary film, *Africa: The Serengeti*. While having the appearance of a frenzy, recent research has shown that a herd of wildebeest possesses what is known as "swarm intelligence", whereby the animals systematically explore and overcome the obstacle as one. Major predators that feed on wildebeest include the lion, hyena, cheetah, leopard, and crocodile, which seem to favor the wildebeest. Wildebeest migrations

Blue Wildebeest *(Connochaetes taurinus)*
(Photo: Rus Koorts) [8]

7   Wikipedia contributors. "Wildebeest." Wikipedia, The Free Encyclopedia. Wikipedia, The Free Encyclopedia, 1 Aug. 2011. Web. 6 Aug. 2011.

8   This image is a reproduction of the photograph taken by Rus Koorts, from Pretoria, South Africa and used with permission. Originally placed on www.flickr.com, the original photograph can be viewed at: http://www.flickr.com/photos/ruslou/1090297277/.

are closely followed by vultures, as Wildebeest carcasses are an important source of food for these scavengers. The vultures consume about seventy percent of the wildebeest carcasses available."[9]

4.2.2 The Lepidoptera Migrations

Following are a few extracts from the Wikipedia article on the subject. "Lepidoptera migration is a biological phenomenon whereby populations of butterflies or moths migrate over long distances to areas where they cannot settle for long periods of time. By migrating, Lepidoptera species can avoid unfavorable circumstances, including weather, food shortage, or over-population. Like birds, there are Lepidoptera species of which all individuals migrate, but there are also species of which only a subgroup of the individuals migrate. The most famous Lepidopteran migration is that of the Monarch butterfly which migrates from southern Canada to wintering sites in central Mexico. The Danaids in South India are also prominent migrators, between Eastern Ghats and Western Ghats.

An important difference with bird migration is that an individual butterfly or moth usually migrates in one direction, while birds migrate back and forth multiple times within their lifespan. This is due to the short lifespan as an imago (the last stage of development of an insect, when they usually have wings). Species that migrate back and forth usually do so in different generations. There are however, some exceptions:

- The famous migration of the Monarch butterfly in North America. This species migrates back and forth in one generation, though it completes only part of the journey in both directions in that generation. No individual completes the entire journey which is spread over a number of generations. The imago of the last summer generation is born in North America, migrates to Mexico, Florida, or California and stays there for the winter. After the winter it migrates back to the north to reproduce. In a couple of generations, the monarch migrates north to Canada.

- The migration of *Agrotis infusa* in Australia. This species migrates from south-eastern Australia to the Australian Alps in the summer to avoid the heat. After the summer it returns to reproduce."[10]

---

9   Wikipedia contributors. "Wildebeest." Wikipedia, The Free Encyclopedia. Wikipedia, The Free Encyclopedia, 1 Aug. 2011. Web. 6 Aug. 2011.

10   Wikipedia contributors. "Lepidoptera migration." Wikipedia, The Free Encyclopedia. Wikipedia, The Free Encyclopedia, 13 Jul. 2011. Web. 6 Aug. 2011.

4.2.3 The Salmon Runs

Bald Eagle *(Haliaeetus leucocephalus)*
(Photo: GollyGforce) [11]

Salmon is the common name for several species of fish in the family *Salmonidae*. Several other fish in the same family are called "trout." The difference is often said to be that salmon migrate and trout are resident, but this distinction does not strictly hold true. Salmon live along the coasts of both the North Atlantic (one migratory species *Salmo salar*) and Pacific Oceans (approximately a dozen species of the genus *Oncorhynchus*), and have also been introduced into the Great Lakes of North America.

"The salmon run is the time at which salmon swim back up the rivers in which they were born to spawn. All Pacific salmon die after spawning. While most Atlantic salmon die after their first spawn, about five to ten percent (mostly female) return to the sea to feed between spawnings. The annual run is a major event for sport fisherman and grizzly bears. The Bald Eagle *(Haliaeetus leucocephalus)*, has an opportunistic and varied diet, but it feeds mainly on fish. In the Pacific Northwest, spawning trout and salmon provide most of the Bald Eagle's diet." [12, 13]

4.3 Bird Ringing / Banding

Bird ringing or bird banding is a technique extensively used in the study of wild birds. By attaching a small, individually numbered, metal or plastic tag to their legs, various aspects of the bird's life can be studied by the ability to re-find, and by the numbering system, positively identify the same individual later. This can include information on migration,

---

[11] This image is a reproduction of the photograph taken by GollyGforce trying to catch up on commenting and used with permission. Originally placed on www.flickr.com, the original photograph can be viewed at: http://www.flickr.com/photos/see-through-the-eye-of-g/4609703815/in/set-72157623531936897.

[12] Wikipedia contributors. "Salmon." Wikipedia, The Free Encyclopedia. Wikipedia, The Free Encyclopedia, 6 Aug. 2011. Web. 6 Aug. 2011.

[13] Wikipedia contributors. "Salmon run." Wikipedia, The Free Encyclopedia. Wikipedia, The Free Encyclopedia, 11 Apr. 2011. Web. 6 Aug. 2011.

longevity, mortality, population studies, territoriality, feeding behavior, and other aspects that are studied by ornithologists.

The ringing of birds was started by a Danish schoolmaster. Hans Christian Cornelius Mortensen (1856–1921) was a Danish teacher and ornithologist. In 1906 he was a co-founder of the Danish Ornithological Society, and in 1909 he was made a corresponding member of the Hungarian Ornithological Society. He was the first to employ bird ringing for scientific purposes. Mortensen was the first to attach zinc/aluminium rings to the legs of birds. In this case, European Starlings (*Sturnus vulgaris*). His report in 1899 gave birth to the practice of bird ringing for scientific purposes.

"In 1919, the Fish and Wildlife Service took responsibility for the co-ordination of banding in the USA and Canada. Since 1996 the US Bird Banding Laboratory collaborates with Canadian programs and, partners with the North American Banding Council (NABC). The European Union for Bird Ringing (EURING) consolidates ringing data from the various national programs in Europe. In Australia, the Australian Bird and Bat Banding Scheme manages all bird and bat ringing information, while SAFRING manages bird ringing activities in South Africa. Bird ringing in India is managed by the Bombay Natural History Society. The National Center for Bird Conservation CEMAVE coordinates a national scheme for bird ringing in Brazil."[14]

A ringed bird is a potential source of information on a wide range of subjects such as migration, longevity, distribution, and status. As many fledging birders would confirm, "A bird in hand...." has proved to be a valuable educational tool as it enables the student to study the features of the bird at length and in detail.

Here are some of the more spectacular results achieved by means of bird ringing as mentioned by Wikipedia in their article "Bird ringing".[15]

- "An Arctic Tern ringed as a chick not yet able to fly, on the Farne Islands off the Northumberland coast in eastern Britain in summer 1982, reached Melbourne, Australia in October 1982, a sea journey of over 22,000 kilometers in just three months from fledging.
- A Manx Shearwater ringed as an adult (at least five years old), breeding on Copeland Island, Northern Ireland, is currently (2003/2004) the oldest known wild bird in the world: ringed in July 1953, it was retrapped in July 2003, at least fifty-five years old. Other ringing recoveries have shown that Manx Shearwaters migrate

---

14   Wikipedia contributors. "Bird ringing." Wikipedia, The Free Encyclopedia. Wikipedia, The Free Encyclopedia, 1 Aug. 2011. Web. 6 Aug. 2011.

15   Wikipedia contributors. "Bird ringing." Wikipedia, The Free Encyclopedia. Wikipedia, The Free Encyclopedia, 1 Aug. 2011. Web. 6 Aug. 2011.

over 10,000 kilometers to waters off southern Brazil and Argentina in winter, so this bird has covered a minimum of one million kilometers on migration alone (not counting day-to-day fishing trips).

- Another bird nearly as old, breeding on Bardsey Island off Wales was calculated by ornithologist Chris Mead to have flown over eight million kilometers (five million miles) during its life (and this bird was still alive in 2003, having outlived Chris Mead)."[16]

4.4 Anti-strike Units

"A *bird strike*—sometimes called birdstrike, avian ingestion (only if in an engine), bird hit, or BASH (Bird Aircraft Strike Hazard)—is a collision between an airborne animal (usually a bird or a bat) and a man-made vehicle, especially aircraft. The term is also used for bird deaths resulting from collisions with man made structures such as power lines, towers and wind turbines."[17]

On the 17th of December 1903 at Kitty Hawk, North Carolina, the brothers Orville and Wilbur Wright made the first flights by man in a power-driven airplane. According to the Wright Brothers' diaries Orville hit a bird in 1905, and since then man began to compete with the birds for airspace. When a bird and an aircraft collide in mid-air, a so-called bird strike, it has serious consequences for both parties. In the thirty years before 2000, bird strikes have caused the death of forty-one pilots and resulted in one hundred and thirty fighter aircraft from ten Western air forces crashing. In addition, airport grounds are vast, open areas, attractive to many species. To ensure the highest level of air safety, air strikes have to be prevented as far as possible, and when they do occur they need to be properly recorded and monitored. This is achieved by applying environmentally sensitive management techniques of various kinds.

Donna Rosenthal, author of *The Israelis: Ordinary People in an Extraordinary Land*, says that Israel has the highest concentration of birds per square meter of any country in the world, and that it also has the world's greatest concentration of fighter aircraft in relation to land area. In recent times, the Israeli Air Force has lost more planes to bird strikes than through enemy actions. This included 1,282 strikes with fighter jets, 696 with helicopters, and 637 with transport planes, since 1972. Because of the high altitudes at which

---

[16] Wikipedia contributors. "Bird ringing." Wikipedia, The Free Encyclopedia. Wikipedia, The Free Encyclopedia, 1 Aug. 2011. Web. 6 Aug. 2011.

[17] Wikipedia contributors. "Bird strike." Wikipedia, The Free Encyclopedia. Wikipedia, The Free Encyclopedia, 11 Jul. 2011. Web. 6 Aug. 2011.

some species, especially the soarers, migrate, and because some species e.g. the Swifts *(Apodidae),* migrate at night too, it is difficult to state exact figures regarding the biannual migrations over the Middle East. It is estimated that the spring migration of raptors over Israel have reached about one and a half million birds of thirty-five species. (See par. 4.2).

> "Eagles may soar, but weasels don't get sucked into jet engines."
> **John Benfield**

It is, therefore, not surprising that the Israelis are considered the leaders in this field. More specifically, there is Professor Yossi Leshem from the University of Tel Aviv, who has been successful in reducing the number of strikes in Israel by eighty-eight percent since 1984. He was the first to do his research by means of *drones*, as the unmanned aircraft he uses to follow and record flock flight patterns, are called. Because he believes that "migrating birds know no political borders" he opened the International Centre for the study of Bird Migration, at Latrun, near Jerusalem. The center's field school, radar facility, and museum of flight safety are attracting researchers from all over the globe, and serving as an educational tool for all the people of the region, involving school children from a number of neighboring countries. Says Leshem: "It's an important symbol for the new Middle East. Instead of doves, maybe storks will bring us peace."

4.5 Extinction

Extinction, noun: extinguishing; making, being, becoming extinct; abolition, wiping out, annihilation.

Of the more than nine thousand bird species, over one thousand are threatened with extinction, and the numbers of five thousand species are declining today. BirdLife International classifies threatened species into one of three categories, based on perceived risk of extinction:

- Critically endangered: fifty percent probability of extinction within five years.
- Endangered: twenty percent probability of extinction within twenty years.
- Vulnerable: ten percent probability of extinction within one hundred years.

Many species of birds have been utilized by man throughout the ages as a relatively easily acquired source of high protein. The Pigeons and Doves *(Columbidae)*, and the

The Dodo *(Raphus cucullatus)*
(Photo: Joel VanAtta) [18]

Fowl-like birds *(Galliformes)* immediately come to mind. This practice is not, as one might have thought, restricted to the more underdeveloped countries. For example: Between 1983 and 1987 the American Mourning Dove *(Zenaida macroura)* had been harvested at an average of almost forty-six million birds per annum. In North-East Brazil, Eared Doves *(Zenaida auriculata)* are harvested at a rate of one hundred thousand per week during the migratory season.

Another source of protein, in this case for the sailors and travelers of the Indian Ocean, were the now extinct family of the Dodos and Solitaires *(Raphidae)*. The visitors found them early in the sixteenth century on the Mascarene Islands of the Southwestern Indian Ocean. Two factors made them particularly susceptible to human predation. Firstly, they were flightless, and could not simply fly away like other birds when threatened. But secondly, they were so unfamiliar with humans, and therefore fearless to the visitors, that they did not feel threatened at all.

They possessed the following characteristics:

- They were large birds, about the size of a Turkey *(Meleagrididae)*.
- They had heavy fat bodies, the males weighing up to twenty-eight kilograms.
- They had exceptionally large heads.
- They had bulbous hooked beaks.
- They had stout strong legs.
- They had ridiculously short wings.
- They had tails of loose, curly feathers.

There were three species: The Dodo *(Raphus cucullatus)*, from Mauritius, which survived until about 1680; the Rodrigues Solitaire *(Pezophaps solitaria)* surviving until almost 1800; and the Reunion Solitaire *(Ornithaptera solitaria)* surviving up to about 1750. In

---

18   This image is a reproduction of the photograph taken by Joel VanAtta and used with permission. Originally placed on www.flickr.com, the original photograph can be viewed at: http://www.flickr.com/photos/joelvanatta/238940638/.

addition, the sailors introduced invaders like pigs and monkeys to the islands. The invaders preyed on the eggs and chicks of the birds and eventually exterminated them.

A suitable habitat provides birds with the three things necessary for survival namely:

- Suitable nesting sites.
- The regular supply of suitable food in sufficient quantities.
- Protection against predation by other beasts and man.

The largest numbers of the Typical Owls *(Strigidae)* inhabit areas of tropical forest, and these are being exterminated at an alarming rate. In addition, many species are endemic to islands and have small population sizes and restricted geographical ranges. All of the above make them extremely vulnerable to habitat alteration, resulting in the destruction of scarce nesting sites, the reduction in food availability, and increasing predation risks. The above is confirmed by the fact that each of the taxa becoming extinct in the last three hundred years, inhabited islands. For example, the Rodrigues Little Owl *(Athene murivora)*, the Mauritius Scops-owl *(Otus commersoni)*, last seen in 1837, and the Mauritius Owl *(Strix sauzieri)* from the islands in the Indian Ocean, had become extinct before scientists could describe them from living specimens. More examples are found in Hawaii, New Zealand, and the West Indies. Four Strigid species are considered Critically Endangered: the Seychelles Scops-owl *(Otus insularis),* the Anjouan Scops-owl *(Otus cepnodes),* the Comoro Scops-owl *(Otus pauliani),* and the Forest Owlet *(Athene blewitti).*

This trend is found in the Pigeons and Doves *(Columbidae)* as well. Nine of the ten species known to disappear into oblivion, the Passenger Pigeon *(Ectopistes migratorius),* being the exception, within the last two centuries were from islands. Of the three hundred and nine spe-

Passenger pigeon *(Ectopistes migratorius)*
(Photo: Wikimedia Commons) [19]

---

[19] This image is a file from the Wikimedia Commons. Full details of the file can be found at: http://en.wikipedia.org/wiki/File:Ectopistes_migratoriusFCN2P29CA.jpg. This image (or other media file) is in the public domain because its copyright has expired. This applies to Australia, the European Union and those countries with a copyright term of life of the author plus 70 years.

cies of pigeons, fifty-eight are currently listed as "Threatened", and forty-seven of these are insular (inhabitant of an island) species.

An article appeared in *Africa Birds and Birding*, Vol. 6, No 2, April/May 2001, in which the author Phil Hockey under the title *African Island Extinctions* provide more detailed information on the above-mentioned. He concludes his article by saying that what we can hope, however, is that the losses of the past together with the danger signals of the present will serve to help us prevent any further damage to the birds and other wildlife of Africa's fascinating but fragile island satellites.

The bird-nest trade is a highly lucrative business, and as a result of over-harvesting the populations of South-East Asian Swiftlets *(Plethora)* have dramatically diminished in recent years. The most famous pigeon in the New World, the Passenger Pigeon *(Ectopistes migratorius)* became extinct in 1914. Only forty-three years earlier a concentration of one hundred and thirty-six million of them bred in an area of two thousand two hundred square kilometers in Wisconsin. In that particular year, six hundred professional hunters harvested more than one million birds. Sixty-one years before that, in other words one hundred and four years before the last specimen on the planet died, Alexander Wilson estimated as many as two billion birds in a single flock.

At Villa Ascasubi in Argentina, Eared Doves *(Zenaida auriculata)* feed on cultivated crops of sorghum, wheat, and millet. In a single treatment of strychnine baits, four hundred and twenty thousand of a population of three million was killed. The destruction of large areas of eastern forest meant fewer patches of mast (the fruit of forest trees, such as beech, oak, etc.) to feed on, and smaller flocks made the detection of rich areas more difficult for passing flocks. This in turn results in large numbers starving to death or at least not being able to reproduce. The tragedy is more heartbreaking if one realizes that the family is well suited for domestication and sustainable harvesting. This can be done in the wild to control numbers of species known to develop into such numbers that they either become a pest or become an environmental problem themselves. In general, the domestication of a species would facilitate the harvesting process, and the savings gained could compensate for the purchase of feed. Grain eating species are particularly suitable as their dietary requirements make them easy to care for.

Conservationists and their organizations cannot compete with the financial power of the capitalist system. One of the solutions would be what I would call: "Meet the Greed." More politely we name it with terms like "habitat preservation", "sustainable harvesting", and "acceptable co-existence", etc.

Now it is becoming clear that the single most important factor leading to the reduction in avian populations is the destruction of their habitat, which is almost exclusively attributable to humans. Intense grazing by domestic or feral (changed from being domesticated to being wild or untamed again) livestock, deforestation for agricultural or other uses, the introduction of exotic plants and trees, the draining of wetlands, global warming, etc, are all reducing and/or altering habitats world wide at an alarming rate. This realization has led to the declaration of Important Bird Areas (IBA's) on a worldwide scale. This concept represents pro-active conservation at its best. Instead of reacting as soon as the developers' earthmoving equipment start moving in, an area is identified using criteria drawn up by BirdLife International and listed as an IBA.

4.6 Nomenclature

Nomenclature is a term that is used to describe a) a list of names and/or terms, or b) the system of principles, procedures and terms related to naming something. In other words the assigning of a word or phrase to a particular object. For example: those names and/or terms used in a particular science or art, by an individual or community.

4.6.1 The Binominal System

As we have seen in previous paragraphs, Aristotle was the first person to think in terms of the classification of all living creatures, (see par. 1.3), but it was Carl von Linne who invented the binominal nomenclature (Latin: *bi* means "two"; *nomen* means "name", thus: pertaining to two names; Latin: *calare* means "to name", thus: "system of names"), (see par. 2.3). Nomenclature is the result of taxonomy (biological classification). The Taxonomist would study the biological characteristics, and based on the information gained, classify the species.

Today, most modern Taxonomists agree that the classification should be based on inferred evolutionary relationships among living things, or evolutionary "trees" called *cladograms*. In recent times, the DNA (deoxyribonucleic acid) structure of a species would be analyzed to confirm or contradict the position of a species. For example: Birds belong to the subphylum *Vertebrata*, as they posses a bony skeleton, a skull and two pairs of limbs. There are no less than twenty-eight categories in use today in the system, of which six are the best known. From most generic (general) to most specific they are: kingdom, phylum, class, order, family, genus, and species. Refer to Chapters 30 and 31 for examples of the system.

The names use Latin grammatical forms, although they can be based on words from other languages. This makes the name universally acceptable, without the danger of being misunderstood by people from differing languages, cultures, backgrounds or geographical areas. The binominal name is given in brackets, in an italic print, after the name used in a particular language. The genus is placed first, followed by the species, and the subspecies, for example: in English: Spotted Eagle-owl *(Bubo africanus milesi)*, in German: Fleck-enuhu *(Bubo africanus milesi)*.

4.6.2 The Living World

In 1959, R.H. Whittaker proposed a five-kingdom system of classifying the entire living world. Refer to Chapter 30 for an example of how a species, in this case, the Spotted Eagle-owl *(Bubo africanus),* would fit into such a system.

- The Kingdom *Protista* consists of the unicellular algae and the protozoans.
- The Kingdom *Fungi* consists of the non-photosynthetic plantlike organisms.
- The Kingdom *Bacteria* can be disease forming like typhoid fever or malaria bacilli, or wholesome like the lactic acid microbes which bring about fermentation.
- The Kingdom *Plantae* is well known.
- The Kingdom *Animalia* consists amongst others of:
  The Phylum *Arthropoda:* 1 000 000 species.
  The Subphylum *Invertebrata:* 232 000 species.
  The Subphylum *Vertebrata* in which the following classes are found:
    - The Class *Pisces:*   20 000 species.
    - The Class *Aves:*   9 000 species.
    - The Class *Reptilia:*   6 000 species.
    - The Class *Mamilia:*   5 000 species.
    - The Class *Amphibia:* 1 500 species.

4.6.3 The Class: Aves

The Class *Aves* is known as the "birds." They are animals that possess the following features:

- Feathered: As apposed to the hair on mammals.
- Winged: Wings are evolved forelimbs.
- Bipedal: Locomotion by means of its two rear limbs, or legs.
- Endothermic: Warm-blooded.
- Egg-laying: They lay hard-shelled eggs.
- Vertebrate: They possess backbones and spinal columns.
  In addition they have:

- An extremely high metabolic rate, a four-chambered heart, a scull, and a lightweight but strong skeleton.
- All living species of birds and most bird species can fly, with some exceptions, including ratites, penguins, and a number of diverse endemic island species.
- Birds also have unique digestive and respiratory systems that are highly adapted for flight.

Some birds, especially Corvids (crows, ravens, etc.) and parrots, are among the most intelligent animal species. A number of bird species have been seen to manufacture and use tools. Many social species exhibit cultural transmission of knowledge across generations. Refer to Chapter 31 for an example of how a particular bird, in this case the Spotted Eagle-owl *(Bubo africanus),* would fit into the binominal system.

4.7 Taxonomy

Taxonomy, noun: from the Ancient Greek: *taxis* which means "arrangement", and *nomia* which means "method", is the practice and science of identifying, classifying, and naming organisms according to their established natural relationships with other organisms. Taxonomy uses taxonomic units known as *taxa* (singular: *taxon*).

Mute Swan *(Cygnus olor)*
(Photo: Ted van den Bergh) [20]

---

[20]   This image is a reproduction of the photograph taken by Ted van den Bergh and used with permission. Originally placed on www.flickr.com, the original photograph can be viewed at: http://www.flickr.com/photos/webted/454540989/in/set-72157605244874930.

Taxonomy itself is never regulated by anyone, but is always the result of research done in the scientific community. The individual researchers arrive at their taxa based on the available data they have gathered through their research, and by utilizing the resources they have chosen to employ in the process. The methods they use to finally name taxa vary considerably. Some would simply make quantitative or qualitative comparisons of some striking features found in a number of individuals which they believe form a taxon. Others for example, use elaborate computer analyses of large amounts of DNA sequenced data.

DNA profiling (also called DNA testing, DNA typing, or genetic fingerprinting) is a technique employed by forensic scientists to assist in the identification of individuals by their respective DNA profiles. Rosalind Franklin, a physical chemist working with Maurice Wilkins at King's College in London, was among the first to use this method to analyze genetic material. Her experiments produced what were referred to at the time as "the most beautiful X-ray photographs of any substance ever taken." The individuals most commonly associated with the discovery of the structure of DNA in 1953 are James Watson and Francis Crick. Maurice Wilkins played a role as well, for which he shared the 1962 Nobel Prize for Physiology and Medicine "for their discoveries concerning the molecular structure of nucleic acids and its significance for information transfer in living material." DNA profiling has "thrown the cat amongst the pigeons", resulting in the taxonomy of birds being in a fluid state at the moment.

For some interesting reading, refer to the text of the Wikipedia article "List of Birds",[21] showing the two Superorders, all the orders, and all the families in which birds are usually placed:

NOTE: The "Old World" refers to the Eastern hemisphere, including Africa, Europe, and Asia.

               The "New World" refers to the Western hemisphere, including the Americas.

4.8 Palestinian Zoogeography

According to the present political borders, the area called Palestine in ancient times can be described as follows: From south Lebanon, Mount Hermon and south-west Syria (including the Golan Heights) in the north, to the Red Sea in the south. From the Mediterranean coast in the west, to the strip of Jordan between the Jordan River

---

[21] Wikipedia contributors. "List of birds." *Wikipedia, The Free Encyclopedia*. Wikipedia, The Free Encyclopedia, 17 Nov. 2011. Web. 1 Dec. 2011.

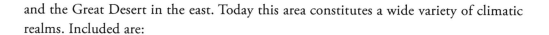

and the Great Desert in the east. Today this area constitutes a wide variety of climatic realms. Included are:

- The subtropical Mediterranean coastal lowlands.
- The Alpine zone of Mount Hermon, snow covered for three to five months of the year.
- The Steppe woodland on the lower slopes of Mount Hermon.
- The moist open grasslands of the Golan Heights.
- The freshwater wetlands of the Hula Valley and the Sea of Galilee.
- The sandy or stony plains of the northern Negev, which is an extension of the Mediterranean maritime plain, with low sparse vegetation, more reminiscent of semi-desert than true desert.
- The central mountain range stretching over the entire length of the country, constituting the largest part of Israel's proper deserts as in Judea and bordering the Syrio-African Rift Valley. Deep wadis running down to the Rift Valley form steep cliffs, and occasional snowfalls occur on the higher ground, e.g. Jerusalem.
- The Syrio-African Rift Valley here runs from Eilat in the south, through the Arava valley, and the Dead Sea depression, on to the Sea of Galilee in the north.

In avian terms, Palestine is situated at the crossroads of three continents namely: Europe, Asia and Africa, and forms a land bridge between the western Palearctic and Afrotropical zoogeographical regions. The Syrio-African Rift Valley starts in the Drakensberg of South-Africa, stretching up through the great lakes area, through north-east Africa, over the Red Sea, up the Gulf of Aquaba, up through the entire Palestine following the Jordan valley and continuing northwards to Turkey. To soaring migrants this constitutes the ideal situation as the Rift Valley ensures a constant supply of thermals all along the way from the breeding areas in the north to the wintering areas in the south, and back.

Some large, broad-winged birds rely on thermal columns of rising hot air to enable them to soar. These include many birds of prey such as vultures, eagles, buzzards, and also storks. Migratory species in these groups have great difficulty crossing large bodies of water, since thermals only form over land, and these birds cannot maintain active flight for long distances. The Mediterranean and other seas present a major obstacle to soaring birds, which must cross at the narrowest water bodies. The dramatic outline of the Rift Valley serves as a corridor or migratory bottleneck for millions of migrants every season with massive numbers passing through Eilat in the southern tip of Israel at migration times. More common species, such as the Honey Buzzard *(Pernis apivorus)* can be

counted in hundreds of thousands in autumn. For this reason the *Eilat Spring Migration Festival* has become a prominent annual feature in birding circles.

Hooded crow *(Corvus cornix)*
(Photo: Ted van den Bergh) [22]

---

[22] This image is a reproduction of the photograph taken by Ted van den Bergh and used with permission. Originally placed on www.flickr.com, under the username "webted", the original photograph can be viewed at: http://www.flickr.com/photos/webted/445159012/.

BIRDS AND BIBLES
IN HISTORY

PART 2
BIBLE TRADITIONS

A concise overview of the processes by which
the different Bibles developed throughout history.

Burrowing Owl *(Athene cunicularia)*
(Photo: Dario Sanches) [1]

CHAPTER 5
BIBLES IN HISTORY

> Moses said to God,
> "Suppose I go to the Israelites and say to them,
>
> 'The God of your fathers has sent me to you,' and they ask me,
> 'What is his name?'
> Then what shall I tell them?"
> God said to Moses, "I AM WHO I AM."
> This is what you are to say to the Israelites: 'I AM' has sent me to you."
>
> **Exodus 3:13–14 (NIV)**[2]

5.1 Chronology

AD or A.D.:	"Anno Domini." Latin for: "In the Year of (Our) Lord."
CE:	"Common Era", "Christian Era", or "Current Era."
BC or B.C.:	"Before Christ."
BCE:	"Before the Common Era", "Before the Christian Era", or "Before the Current Era."

The so-called "Christian calendar", which is in fact the Gregorian calendar, is based on the traditionally reckoned year of the birth of Jesus of Nazareth. This AD designation originating among Christians in Europe at least as early as 1615 (at first in Latin), but in the light of new evidence, it is now generally accepted that the actual date of his birth was somewhere around 4 BC. AD counts the years after the start of this epoch, and BC denoting years before the start of the epoch. There is no year zero in this scheme, so the year 1 AD immediately follows the year 1 BC.

The Gregorian calendar also known as the Western calendar, or Christian calendar, is internationally the most widely used civil calendar in the world today. For decades, it has

---

[2]   THE HOLY BIBLE, NEW INTERNATIONAL VERSION®, NIV® Copyright © 1973, 1978, 1984, 2011 by Biblica, Inc.™ Used by permission. All rights reserved worldwide.

been the unofficial global standard, recognized by international institutions such as the United Nations and the Universal Postal Union. Though the Western calendars spread around the world over the past four centuries as a result of western power, in modern times the Gregorian calendar and AD year numbering has been adopted by many non-Western countries with no Christian heritage.

CE is often preferred by those who desire to employ a term that is not directly and explicitly related to an event prominent in the Christian faith. For example: Upon its foundation in 1911, the Republic of China adopted the Western calendar in 1912 and the translated term was *xi yuan*, meaning "western era." Later, with the founding of the People's Republic of China in 1949, it reiterated the use of the Gregorian calendar and accepted the term *gong yuan*, which means "common / public era."

Until about a century ago, Old Testament dates were calculated almost entirely from the biblical statements and narrations. However, the Old Testament does not provide sufficient details, and ancient versions like the *Septuagint* (See par. 5.4.2), sometimes offer variant figures. Today scholars like historians correlate data from a wider spectrum of sources. For example, good dates are available based on Egyptian data from ca. 2100 BC onwards, and Mesopotamia data from ca. 1400 BC onwards. Within the scope of this book it is impossible to discuss the vast amounts of documentation, which underlie the dates.

5.2 The Holy Bible

What laymen today call "The Bible" (Greek: *ta biblia* "the books"), also known as "The Holy Bible", are collections (not only one) of sacred scriptures related to Judaism and Christianity. Bibles have always been the best-selling books in history with an estimated more than six billion copies published to date. The word "Bible" comes from the Greek *biblios*, meaning "book"; and the word "manuscript" comes from the Latin *manu*, meaning "hand", and *scriptum*, meaning "written."

There is no single book that can be called "The Bible", (see par. 6.1–4 for an explanation), as the individual books, their content, and the order of the books vary between different religions and even denominations. For example: mainstream Judaism divides the *Tanakh* (the canon of the Hebrew Bible, see par. 5.3) into twenty-two books, but they are divided into thirty-nine books in the Christian Old Testament.

The Old Testament is accepted by Christians as Scripture (the texts which various religious traditions consider to be sacred). Broadly speaking, it contains the same material as the Hebrew Bibles. However, the order and number of the books is not entirely the same as that found in Hebrew manuscripts and in the ancient versions.

5.3 The Jewish Bible

The *Tanakh* is a name used in Judaism for the canon of the Hebrew Bibles. The name is an acronym formed from the initial Hebrew letters of the Masoretic Text's three traditional subdivisions:

- The *Torah* ("Teaching"), also known as "The Five Books of Moses", or the *Pentateuch* in Greek meaning "five scroll-cases." It comprises narratives about the origins of the Israelite nation, its laws, and its covenant with the God of Israel.
- The *Nevi'im* ("Prophets"), containing the historic account of the ancient kingdoms of Israel and Judah, as well as an account of the lives, work, and prophecies of the so-called "large prophets", and the so-called "small prophets."
- The *Ketuvim* ("Writings"), are poetic and philosophical works such as the Psalms, Proverbs, and the Book of Job.

Therefore, the acronym *Tanakh* came into being. The books of the *Tanakh* were relayed with an accompanying oral tradition passed on by each generation, called the "*Oral Torah*." The thirty-nine books found in the Old Testament of today's Christian Bibles are grouped together in the *Tanakh* as twenty-two books, equaling the number of letters in the Hebrew alphabet. This canon of Jewish scripture is attested to (declare to be correct) by at least four sources, namely:

A passage from the *Tanakh*
(*Photo: Wikimedia Commons*) [3]

- Philo (20 BC–50 AD), known also as Philo of Alexandria.
- Titus Flavius Josephus (37–100 A.D.).
- The New Testament (Luke 24:44).
- The *Talmud* (a central text of mainstream Judaism).

---

[3]   This image is a file from the Wikimedia Commons. Full details of the file can be found at: http://en.wikipedia.org/wiki/File:Targum.jpg. This image (or other media file) is in the public domain because its copyright has expired. This applies to Australia, the European Union and those countries with a copyright term of life of the author plus 70 years.

The Bibles used at Qumran, home to a conservative Jewish sect, the Essenes, on the banks of the Dead Sea, excluded the book Esther but included a book called Tobit. (See par. 5.4.5). Otherwise, it seems to have been basically the same as the Hebrew Bibles or Old Testament, albeit with many textual variants.

According to the *Talmud*, much of the contents of the *Tanakh* was compiled by the "Men of the Great Assembly" by 450 BC, and has since remained unchanged. Modern scholars believe that the process of canonization of the *Tanakh* became finalized between 200 BC and 200 AD.

The Hebrew text was originally an *abjad*: consonants written with some applied vowel letters the so-called "*matres lectionis.*" Later, during the early Middle Ages, scholars known as the Masoretes created a single formalized system of vocalization. This was chiefly done by Aaron ben Moses ben Asher, in the Tiberias School, based on the oral tradition for reading the *Tanakh*, hence the name Tiberian vocalization.

5.3.1 The *Torah*

In its most limited sense, *Torah* refers to the Five Books of Moses: Genesis, Exodus, Leviticus, Numbers and Deuteronomy. But the word *Torah* can also be used to refer to the entire Jewish Bible, or in its broadest sense, to the whole body of Jewish law and teachings. The *Torah* also known as the *Pentateuch* (Greek: *penta-* "five", and *teuchos* "tool, vessel, book"), focuses on three aspects in the changing relationship between God and the Hebrew people.

The first eleven chapters of Genesis provide accounts of the creation (or ordering) of the world and the history of God's early relationship with humanity. The remaining thirty-nine chapters of Genesis provide an account of God's covenant with the Hebrew patriarchs, Abraham, Isaac and Jacob (also called Israel), and Jacob's children (the "Children of Israel"), especially Joseph. It tells of how God commanded Abraham to leave his family and home in the city of Ur, eventually to settle in the land of Canaan, and how the "Children of Israel" later moved to Egypt. The remaining four books of the *Torah* relate the story of Moses, who lived hundreds of years after the patriarchs. His story coincides with the story of the liberation of the "Children of Israel" from slavery in Ancient Egypt, to the renewal of their covenant with God at Mount Sinai, and their wanderings in the desert until a new generation would be ready to enter the land of Canaan. The *Torah* ends with the death of Moses.

The *Torah* contains the commandments of God, revealed at Mount Sinai (although there is some debate amongst Jewish scholars as to whether this was written down completely in one moment, or if it was spread out during the forty years in the wandering in the desert). These commandments provide the basis for *Halakha* (Jewish religious law). Tradition states that the number of these is equal to six hundred and thirteen commandments, which are divided into three hundred and sixty-five restrictions and two hundred and forty-eight positive commands. The *Torah* is divided into fifty-four sections which are read on successive Sabbaths in Jewish liturgy, from the beginning of Genesis to the end of Deuteronomy.

5.3.2 The *Nevi'im*

The *Nevi'im*, or "Prophets", relate the story of the rise of the Hebrew monarchy, its eventual division into two kingdoms (Israel in the north, and Judah in the south), and the prophets who, in God's name, warned the kings and the "Children of Israel" about the punishment of God whenever they would stray and worship gods other than Jahweh. It ends with the conquest of the Kingdom of Israel by the Assyrians and the conquest of the Kingdom of Judah by the Babylonians, and the destruction of the Temple in Jerusalem. Portions of the prophetic books are read by Jews on the Sabbath. The Book of Jonah is read on The Day of Atonement (*Yom Kippur*). According to Jewish tradition, the *Nevi'im* is divided into eight books. Contemporary translations subdivide these into twenty-one books.

5.3.3 The *Ketuvim*

The *Ketuvim*, or "Writings" or "Scriptures", may have been written during or after the Babylonian Exile. Many of the psalms in the book of Psalms are attributed to King David. King Solomon is believed to have written Song of Songs in his youth, Proverbs during the prime of his life, and Ecclesiastes in his old age. The prophet Jeremiah is thought to have written Lamentations. The Book of Ruth is the only biblical book that centers entirely on a non-Jew. It tells the story of a non-Jewish lady from Moab who married a Jew and, upon her husband's death, followed in the ways of the Jews. According to the Bibles, she was the great-grandmother of King David. Five of the books, called "The Five Scrolls" (Hebrew: *Megillot*), are read on Jewish holidays: Song of Songs on Passover; the Book of Ruth on Shavuot; Lamentations on the Ninth of Av; Ecclesiastes on Sukkot; and the Book of Esther on Purim. Collectively, the Ketuvim contain lyrical poetry, philosophical reflections on life, and the stories of the prophets and other Jewish leaders during the Babylonian exile. It ends with the Persian decree called the *Edict of Restoration* allowing Jews to return to Jerusalem to rebuild the Temple.

Christian Bibles include the books of the Hebrew Bibles, but arranged in a different order: Jewish Scripture ends with the people of Israel restored to Jerusalem and the temple, but the Christian arrangement ends with the book of the prophet Malachi.

> "People expect the clergy to have the grace of a swan, the friendliness of a sparrow, the strength of an eagle, and the night hours of an owl – and some people expect such a bird to live on the food of a canary."
>
> **Edward Jeffrey**

5.4 The Ancient Versions

Modern translations are made from the available ancient texts called "versions." These ancient translations of the Hebrew texts were made either by Jewish communities no longer familiar with Hebrew, or by Christians whose culture and languages were quite distinct from Judaism and Hebrew. Here we shall look at the versions in chronological order rather than in order of importance. In a nutshell one can say that the Old Testament was translated into three languages, namely:

- The *Targums*: into Aramaic.
- The *Septuagint*: into Greek.
- The *Vulgate*: into Latin.

5.4.1 The Targums

From about the tenth century BC, the Aramaic language was widely spread in northern Palestine and Syria (Aram). By the time of the Persian Empire, it had become the official language of the western provinces of that Empire. By the time of the prophet Nehemiah, Hebrew was already declining in popular circles, and that is why the book of Daniel, intended as a popular work, has a large section in Aramaic. The book is usually dated about 168–165 BC. By the time of Christ, Aramaic was the most widely spoken language of the common people of Palestine and surroundings. However, documents such as the Dead Sea Scrolls, as well as the Masoretic Texts itself, are a clear indication of the preservation of Hebrew among the learned scholars and more so in synagogue worship.

Septuagint passage
(Photo: Wikimedia Commons) [4]

By the sixth century BC, Aramaic translations of the Hebrew Bibles were required by the Jewry in Palestine, enabling them to learn about their faith and their history in the language they were familiar with at that time. Initially the writing of such translations was forbidden, and only oral translations by the official synagogue translator were allowed. It was done verse by verse in the case of the Law, and every three verses in the case of the Prophets. These translations were known as the *Targums* (in Hebrew: *targum* means "translation, interpretation"). The values of the *Targums* are in the fact that they simultaneously were translations, explanations and sermons, often solving misinterpretations. Some of the best known *Targums* are by Onkelos and Jonathan, dated in the fifth century BC.

5.4.2 The Septuagint

The *Septuagint* is the most important of the versions for four reasons namely:

- It is the oldest.
- It witnesses an ancient Hebrew text.
- It is an early translation into a non-Semitic language.
- It was the Old Testament of the Early Church.

The history of the *Septuagint* is complicated and stretches over a very long period. What follows is a brief description of its origin. After Alexander the Great's death in 323

---

[4] This image is a file from the Wikimedia Commons. Full details of the file can be found at: http://en.wikipedia.org/wiki/File:Codex_vaticanus.jpg. This image (or other media file) is in the public domain because its copyright has expired. This applies to Australia, the European Union and those countries with a copyright term of life of the author plus 70 years.

BC, Egypt was ruled by the Ptolemies. Aramaic was replaced by Greek as the official language, and by the third century BC the large number of Jews living in Alexandria were Greek speaking. The need for a translation of the Hebrew into Greek for use in the synagogues of such Greek-speaking Jews was eminent. Legend has it that Ptolemy II requested the Jerusalem High Priest to authorize a translation of the Pentateuch into Greek for his library. Seventy-two (Latin: *septuaginta,* "seventy") Jewish scholars were sent from Jerusalem and in seventy-two days they emerged with a unified translation, to this day called the *Septuagint.* This translation was given added importance by the promulgation and spread of Christianity in the first century AD. However, the Greek-speaking Jews were dissatisfied with it and required improved translations, which led to the versions of Aquila, Symmachus and Theodotion.

5.4.3 The Vulgate

The need for a Latin translation of the Bibles, especially in North Africa, arose as Christianity spread westwards through the Roman Empire in the early centuries AD. The first Latin translations were then indeed made in North Africa from the *Septuagint,* at that time still regarded as the Bible of the Church. These so-called "Old Latin" versions are witnesses to the *Septuagint* rather than the Hebrew. In 382 AD, Pope Damascus requested the scholar Jerome to revise the current Latin Bible. He translated the New Testament and Psalms, resulting in the Roman Psalter (a book containing the Book of Psalms, often with other devotional material), which is used to this day in St Peter's, Rome. Jerome visited Bethlehem and became convinced that the Hebrew was the true Old Testament text. He familiarized himself with the Hebrew language and rabbinic methods, and in 405 AD he had completed a translation of the Old Testament called The *Vulgate.* The *Vulgate* is important for, amongst others, the following reasons:

- It marks the first recognition of the Church of the primacy of the Hebrew text.

- It is a witness to the Hebrew text as it was at the end of the fourth century AD, five hundred years before the MasoreticText.

- Because of Jerome's exposure to Judaism, we could safely infer that the *Vulgate* preserved early Jewish traditions, especially concerning the meaning of Hebrew words.

5.4.4 The Masoretic Texts

Modern day translations of the Old Testament depend, amongst others, on manuscripts from the ninth and tenth centuries AD, called the Masoretic Texts. They are the oldest written manifestations of the original Hebrew texts. Together they constitute the earliest official and complete Hebrew Bible at our disposal.

Before the fifteenth century all books were copied by hand. The copyists often made unintentional mistakes, resulting in incorrect texts. The Masoretes, a body of Jewish scholars who worked in Palestine and Babylonia ca. 500–1000 AD, sought to make all the differing manuscripts that they were faced with conform to the ideal of a single, official Jewish Bible. This would then conform to their standards and would also restrict conflicting interpretations that could arise. One of the major problems was that the texts were consonantal, to which the Masoretes added vowel pointings. The vowels aided layman in reading the texts and reduced the possibility of misinterpretations. In addition the Masoretes made intentional changes and/or added footnotes to correct what in their opinion had been text corruptions.

5.4.5 The Dead Sea Scrolls

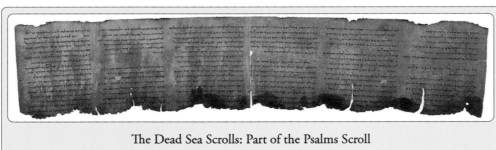

The Dead Sea Scrolls: Part of the Psalms Scroll
(Photo: Wikimedia Commons) [5]

In 1947, while herding goats at *Khirbet Qumran* on the northwest shore of the Dead Sea an Arab boy found a cave. He discovered some vessels filled with scrolls. Today they are known as The Dead Sea Scrolls. They consist of a collection of 972 texts from the Hebrew

---

[5]   This image is a file from the Wikimedia Commons. Full details of the file can be found at: http://en.wikipedia.org/wiki/File:Psalms_Scroll.jpg. This image (or other media file) is in the public domain because its copyright has expired. This applies to Australia, the European Union and those countries with a copyright term of life of the author plus 70 years.

Bible and other extra-biblical documents. The scrolls date back to 150 BC–70 AD. They were written by members of the Essene sect that lived at *Khirbet Qumran* from about 200 BC but had to flee before the Romans in 70 AD. They consist of a complete text of Isaiah and fragments of every other Old Testament book except Esther. They were eleven centuries older than the oldest texts of the time, and confirmed beyond any doubt the general accuracy of the Masoretic Texts. Refer to Chapter 29 for titles on works dealing with this community.

5.5 The Christian Bibles

Christian Bibles are divided into two parts. The first is called the "Old Testament", usually containing the thirty-nine books of Hebrew Scriptures that led to the formation of Judaism. The second part is called the "New Testament", containing twenty-seven books and letters. The first four books of the New Testament form the Canonical gospels which recount the life of Jesus Christ and are central to the Christian faith.

"Testament" is a translation of the Greek *diatheke*, also often translated as "Covenant." It is a legal term denoting a formal and legally binding declaration of benefits to be given by one party to another. For example: "the last will and testament" in secular use. Here it does not connote mutuality (including two equal parties), but rather a unilateral covenant offered by God to humanity. Groups within Christianity include differing books as part of one or both of these "Testaments" of their sacred writing. Many of these differences are because of the inclusion or exclusion of the biblical *Apocrypha* or deuterocanonical books. The general term *Apocrypha* is usually applied to the books that the Protestant Christian Church considers useful but not divinely inspired.

5.5.1 The Old Testament

A "testament" is a document that the author has sworn to be true.

A "covenant" is a solemn agreement to engage in or refrain from a specified action. It is commonly found in religious contexts, where it refers to sacred agreements between God and human beings.

A "communion" is a group of persons having a common religious faith; a religious denomination: For example: the Anglican Communion.

The Old Testament is the collection of books compiled, written, and/or edited by members of the Hebrew-Jewish religious community between approximately the twelfth century BC and the beginning of the Christian era. All the autographs (a document tran-

אַשְׁרֵי־הָאִישׁ
אֲשֶׁר ׀ לֹא הָלַךְ בַּעֲצַת רְשָׁעִים
וּבְדֶרֶךְ חַטָּאִים לֹא עָמָד
וּבְמוֹשַׁב לֵצִים לֹא יָשָׁב:
כִּי אִם בְּתוֹרַת יְהוָה חֶפְצוֹ
וּבְתוֹרָתוֹ יֶהְגֶּה יוֹמָם וָלָיְלָה:

Psalm 1:1 in Biblical Hebrew
(Photo: Wikipedia Commons) [6]

scribed entirely in the handwriting of its author) were lost or destroyed by that time. As soon as a copy had been completed, the original document was destroyed to prevent any possible desecration by enemies in future. The Old Testament forms the first of two parts of the Christian Biblical canon (material accepted as "official and authoritative for faith and teaching"). The content of the Old Testament canon varies from church to church, with the Orthodox communion (also called the Orthodox Catholic Church or the Eastern Orthodox Church) having the most, namely fifty books. Thirty-nine books are shared by all communions. They are the books of the shortest canon, namely that of the major Protestant communions. The differences between the Jewish and Protestant versions are easily explained: The Jewish Bibles contains a book called "The Twelve." It contains twelve prophetic writings which are treated as individual books in Christian versions. The eight books of the Christian version namely: I–II Samuel, I–II Kings, I–II Chronicles, Ezra-Nehemiah, is in turn presented as four books in Jewish Bibles.

The word "testament", Hebrew *berith*, Greek *diatheke*, primarily signifies the covenant which God entered into first with Abraham, and later on Mount Sinai with the "People of Israel." The Prophets had knowledge of a "New Covenant" (Jer 31:31–34) to which the one concluded on Mount Sinai should, according to the Christian doctrine, give way. This is clearly illustrated in the letter to the Hebrews, Chapter 8, in the New Testament. Christ at the Last Supper speaks of "the blood of the New Testament." The Apostle St. Paul declares himself a minister "of the new testament" (2 Corinthians 3:6), and calls the covenant entered into on Mount Sinai "the old testament", (2 Cor 3:6–18; Heb 9:1–4).

The Old Testament includes a wide range of diverse materials some being historical (e.g. I and II Kings), some legalistic (e.g. Leviticus), some poetic (e.g. Psalms), some apocalyptic (e.g. parts of Daniel), and some didactic (e.g. Ecclesiastes). For the most part it was written in Hebrew, but a few passages were written in Aramaic, the *lingua franca* among the Jews during the post-Exilic era (after the sixth century BC). The Aramaic portions include Dan 2:4b–7:28; Ezra 4:8–6:18, 7:12–26; Jer 10:11; and one phrase in Gen 31:47. Much of the material, including many genealogies, poems and narratives, is

---

[6]   This image is a file from the Wikimedia Commons. Full details of the file can be found at: http://en.wikipedia.org/wiki/File:Bhs_psalm1.gif. This file is licensed under the Creative Commons Attribution-Share Alike 3.0 Unported license. A framed copy of this image is reproduced here.

thought to have been handed down by word of mouth for many generations. Very few manuscripts are said to have survived the destruction of Jerusalem in 70 AD.

5.5.2 The New Testament

The New Testament is also an anthology of books and letters, written at different times by various authors. In almost all Christian traditions today, the New Testament consists of twenty-seven books and letters. The original texts were written beginning around 50 AD in Koine (Hellenistic, Common) Greek, the *lingua franca* of the eastern part of the Roman Empire at the time where they were composed. Koine is the main ancestor of Modern Greek. As the language of the New Testament and of the

Papyrus 46
(Photo: Wikimedia Commons) [7]

Church Fathers, Koine Greek is also known as Biblical, Patristic (Latin: *pater*, father), or New Testament Greek. All of the inspired works which would eventually be incorporated into the New Testament would seem to have been written no later than the mid-second century AD.

Parts of the New Testament have been preserved in more manuscripts than any other "ancient" work. Over five thousand eight hundred complete or fragmented Greek manuscripts, ten thousand Latin manuscripts, and nine thousand three hundred manuscripts in various other ancient languages have been preserved. The dates of these manuscripts range from ca. 125 AD to the introduction of printing in Germany in the fifteenth

---

[7]   This image is a file from the Wikimedia Commons. Full details of the file can be found at: http://en.wikipedia.org/wiki/File:P46.jpg. This image (or other media file) is in the public domain because its copyright has expired. This applies to Australia, the European Union and those countries with a copyright term of life of the author plus 70 years.

century. However, the original manuscripts of the New Testament books do not survive today. The autographs (a document transcribed entirely in the handwriting of its author) were lost or destroyed a long time ago. What survives for us to work with today, are copies and transcriptions of the originals. For example: *Papyrus 46*, is one of the oldest extant New Testament manuscripts in Greek at our disposal, written on papyrus, with its "most probable date" between 175-225 AD.

5.5.3 The History of Bible Translations

Previously we have dealt with three ancient translations of the Hebrew Bibles, namely:

- The Aramaic translations,   the *Targums*,   in par. 5.4.1
- The Greek translation,   the *Septuagint*,   in par. 5.4.2
- The Latin translation,   the *Vulgate*,   in par. 5.4.3.

Now let us have a look at the rest of the history, based largely on the comprehensive Wikipedia article "Bible Translations."[8]

5.5.3.1 Early Translations in Late Antiquity

Origen Adamantius, (184/5–253/4), was a Christian scholar and theologian from Alexandria. Origen's *Hexapla* placed six versions of the Old Testament, including second century Greek translations, side by side. The canonical Christian Bible was formally established by Bishop Cyril of Jerusalem in 350 (although it had been generally accepted by the church previously), confirmed by the Council of Laodicea in 363 AD (both lacked the book of Revelation), and later established by Athanasius of Alexandria in 367 AD (with Revelation added). Christian translations also tend to be based upon the Hebrew, though some denominations prefer the *Septuagint* (or may cite variant readings from both). Bible translations incorporating modern textual criticism usually begin with the Masoretic text, but also take into account possible variants from all available ancient versions. The received text of the Christian New Testament is in Koine Greek, and nearly all translations are based upon the Greek text.

---

[8]   Wikipedia contributors. "Bible translations." Wikipedia, The Free Encyclopedia. Wikipedia, The Free Encyclopedia, 28 Jul. 2011. Web. 6 Aug. 2011.

There are also several ancient translations, most important of which are in the Syriac dialect of Aramaic, in the Ethiopian language of Ge'ez, and in Latin. The Bishop of the Goths Wulfila in the today's Bulgaria translated the Bible into Gothic in the mid-4th century.

5.5.3.2 The Middle Ages

The earliest surviving complete manuscript of an entire Bible is the *Codex Amiatinus*, a Latin *Vulgate* edition produced in eight century England at the double monastery of Wearmouth-Jarrow.

During the Middle Ages, translation, particularly of the Old Testament, was discouraged by the Church. Nevertheless, there are some fragmentary Old English Bible translations, notably a lost translation of the Gospel of John into Old English by the Venerable Bede, which he is said to have been prepared shortly before his death around the year 735. An Old High German version of the gospel of Matthew dates to 748 AD. The translation into Old Church Slavonic dates to the late ninth century.

5.5.3.3 Reformation and Early Modern Period

The earliest printed edition of the Greek New Testament appeared in 1516 from the Froben press, by Desiderius Erasmus, who reconstructed its Greek text from several recent manuscripts of the Byzantine text-type. In 1521, Martin Luther was placed under the Ban of the Empire, and he retired to the Wartburg Castle. During his time there, he translated the New Testament from Greek into German. It was printed in September 1522. The first complete Dutch Bible, partly based on the existing portions of Luther's translation, was printed in Antwerp in 1526 by Jacob van Liesvelt.

The use of numbered chapters and verses was not introduced until the Middle Ages and later the system used in English was developed by Stephanus (Robert Estienne of Paris). Early manuscripts of the letters of Paul and other New Testament writings show no punctuation whatsoever. The punctuation was added later by other editors, according to their own understanding of the text.

The churches of the Protestant Reformation translated the Greek of the *Textus Receptus* to produce vernacular Bibles, such as the German Luther Bible, the Polish Brest Bible and the English King James Bible.

Tyndale's New Testament translation (1526, revised in 1534, 1535 and 1536) and his translation of the Pentateuch (1530, 1534) and the Book of Jonah were met with heavy

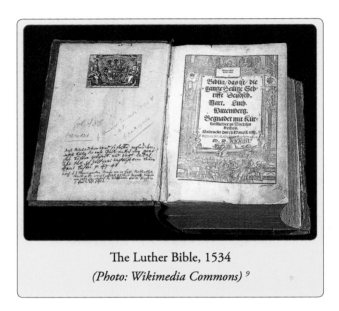

The Luther Bible, 1534
(Photo: Wikimedia Commons) [9]

sanctions given the widespread belief that Tyndale "changed" the Bible as he attempted to translate it. The first complete French Bible was a translation by Jacques Lefevre d'Etaples, published in 1530 in Antwerp. The Froschauer Bible of 1531 and the Luther Bible of 1534 (both appearing in portions throughout the 1520's) were an important part of the Reformation.

The first English translations of Psalms (1530), Isaiah (1531), Proverbs (1533), Ecclesiastes (1533), Jeremiah (1534) and Lamentations (1534), were executed by the Protestant Bible translator George Joye in Antwerp. In 1535 Myles Coverdale published the first complete English Bible also in Antwerp.

In 1584 both Old and New Testaments were translated to Slovene by Protestant writer and theologian Jurij Dalmatin. The Slovenes thus became the twelfth nation in the world with a complete Bible in their language. The missionary activity of the Jesuit order led to a large number of seventeenth century translations into languages of the New World.

Over time a variety of linguistic, philological and ideological approaches to the translation process have been used, including:

- Dynamic equivalence translations. Dynamic equivalence (also known as functional equivalence) attempts to convey the thought expressed in a source text (if necessary, at the expense of literalness, original word order, the source text's grammatical voice, etc.). The term "dynamic equivalence" is associated with the translator

---

[9] This image is a file from the Wikimedia Commons. Full details of the file can be found at: http://en.wikipedia.org/wiki/File:Lutherbibel.jpg. Torsten Schleese, the copyright holder of this work, released this work into the public domain on 14 December 2005. This applies worldwide. In some countries this may not be legally possible; if so: he grants anyone the right to use this work for any purpose, without any conditions, unless such conditions are required by law.

Eugene Nida, and was originally coined to describe ways of translating Bibles, but this approach is applicable to any translation.

- Formal equivalence translations (similar to literal translation). Formal equivalence attempts to render the text word-for-word (if necessary, at the expense of natural expression in the receptor language).
- Idiomatic, or Paraphrastic translations. A paraphrase typically explains or clarifies the text that is being paraphrased. Kenneth N. Taylor (1917–2005), finished an entire Bible in contemporary language and published it as *The Living Bible* in 1971.

English Translations are so numerous that we cannot deal with it at all here. Wikipedia have several related articles. In their article "Bible Translations",[10] they deal with English translations under the following headings:

- English translations of the Bible
- Old English Bible translations
- Middle English Bible translations
- Golden Legend
- Early Modern English Bible translations
- Modern English Bible translations
- Jewish English Bible translations

To this day, Bibles continue to be the most translated books in the world. The following numbers are approximations. As of 2005, at least one Bible book has been translated into 2,400 of the 6,900 languages listed by SIL, including 680 languages in Africa, followed by 590 in Asia, 420 in Oceania, 420 in Latin America and the Caribbean, 210 in Europe, and 75 in North America. The United Bible Societies are presently assisting in over 600 Bible translation projects. Bibles are available in whole or in part to some ninety-eight percent of the world's population in a language in which they are fluent.

The United Bible Society announced that as of 31 December 2007 that Bibles were available in 438 languages, 123 of which included the deuterocanonical material as well as the *Tanakh* and New Testament. Either the *Tanakh* or the New Testament alone was available in an additional 1168 languages, and portions of Bibles were available in another 848 languages, for a total of 2,454 languages.

Translation from one language (the source language) to another (the receptor language) is a daunting task, as every language has its own uniqueness. Translating Bibles is

---

[10] Wikipedia contributors. "Bible translations." Wikipedia, The Free Encyclopedia. Wikipedia, The Free Encyclopedia, 28 Jul. 2011. Web. 6 Aug. 2011.

much more complex as the translators are dealing with ancient works of which they do not even have the original autographs.

Anyone studying the birds in their Bibles very soon realizes the enormity of the task. In writing and compiling this publication I have attempted to deal with it to some extent. The next section will give the reader a concise introduction into the wide field of textual criticism. No serious student of Bibles can afford not to pay attention to this complex but most enlightening field.

Amethyst Sunbird *(Chalcomitra amethystina)*
(Photo: Rus Koorts) [11]

---

[11] This image is a reproduction of the photograph taken by Rus Koorts from Pretoria, in South Africa and used with permission. Originally placed on www.flickr.com, the original photograph can be viewed at: http://www.flickr.com/photos/ruslou/5631002583/in/set-72157625798426776

CHAPTER 6
TEXTUAL CRITICISM

> "The eye that mocks a father and scorns a mother.
> The ravens of the valley will pick it out,
> and the young eagles will eat it."
> *Prv 30:17 (NASB)[1]*

6.1 The Bible?

Textual criticism is a branch of literary criticism that is concerned with the identification and removal of transcription errors (in the widest sense of the word) in the texts of manuscripts. Ancient scribes often made errors or alterations when copying manuscripts by hand. Given one (or more) copies of a manuscript, but not the original document, the textual critic seeks to reconstruct the original text (the "archetype" or "autograph") as closely as possible. The ultimate objective of the textual critic's work is the production of a "critical edition" containing a text most closely approximating the original. Nida and Taber, in their groundbreaking book *The Theory and Practice of Translation* describe it as the process by which the translator is attempting to reproduce in the receptor language the "closest natural equivalent" of the source-language message.

6.1.1 The Time Factor

The original texts of the Old Testament books, as found in Christian Bibles were written over a period spanning more than one thousand years. By the time of Jesus Christ, and even three to four hundred years before him, all of these original documents (the so-called autographs) were lost, with only copies made over centuries being available.

---

[1] Scripture quotations taken from the New American Standard Bible®, Copyright © 1960, 1962, 1963, 1968, 1971, 1972, 1973, 1975, 1977, 1995 by The Lockman Foundation. Used by permission. (www.Lockman.org).

Black-necked Stilt *(Himantopus mexicanus)*
(Photo: Ingrid Taylar) [2]

Jesus, being a commoner, spoke Aramaic and not Hebrew. In his time the available copies of the Old Testament were written in Hebrew consonants, Aramaic, and Greek. Before the discovery of the Dead Sea Scrolls, the oldest Hebrew texts came from the Masoretes. For centuries the oldest complete Hebrew Old Testament was the so-called *Codex Leningradensis,* dated 1008 AD, named after the museum in Leningrad where it is conserved. Then in 1947 the Dead Sea Scrolls were discovered, and although they are not a complete Old Testament, scholars now had texts in hand that were one thousand years older than *Codex Leningradensis,* and in addition the Scrolls serve as a kind of "cross reference", confirming the accuracy of the Masoretic Texts.

6.1.2 The Manuscripts

The *Aleppo Codex* (ca. 920 AD) and *Leningrad Codex* (ca. 1008 AD) are the oldest Hebrew manuscripts of the *Tanakh.* "The *Leningrad Codex* (or *Codex Leningradensis*) is the oldest complete manuscript of the Hebrew Bible in Hebrew, using the Masoretic text and Tiberian vocalization. It is dated 1008/9 AD according to its colophon (brief description). The *Aleppo Codex*, against which the *Leningrad Codex* was corrected, is several decades older, but parts of it have been missing since 1947, making the *Leningrad Codex* the oldest complete codex of the Tiberian Masoretic Texts that has survived intact to this day. In modern times, the *Leningrad Codex* is most important and it is therefore the Hebrew text reproduced in *Biblia Hebraica* (BHK) (1937) and *Biblia Hebraica Stuttgartensia*

---

2   This image is a reproduction of the photograph taken by Ingrid Taylar, Seattle, WA, USA and used with permission. Originally placed on www.flickr.com, the original photograph can be viewed at: http://www.flickr.com/photos/taylar/2177069658/.

(BHS) (1977), and will be used for *Biblia Hebraica Quinta* (BHQ)."[3] The BHK and BHS are easily and cheaply available, and therefore commonly used by biblical Hebrew students and scholars. It is predicted that *Biblia Hebraica Quinta* will be completed by 2015 or 2020.

The 1947 find at Qumran of the Dead Sea scrolls pushed the manuscript history of the *Tanakh* back a millennium from the two earliest complete codices. Before this discovery, the earliest extant manuscripts of the Old Testament were in Greek in manuscripts such as *Codex Vaticanus* and *Codex Sinaiticus*. Out of the roughly eight hundred manuscripts found at Qumran, two hundred and twenty are from the *Tanakh*. Every book of the *Tanakh* is represented except for the Book of Esther; however, most are fragmentary. Notably, there are two scrolls of the Book of Isaiah, one complete (1QIsa), and one around seventy-five percent complete (1QIsb). These manuscripts generally date between 150 BC and 70 AD. The Shrine of the Book (Hebrew: *Heikhal HaSefer*), a wing of the Israel Museum near Givat Ram in Jerusalem, now houses the Dead Sea Scrolls.

6.2 In Conclusion

Thus, we have to concede that, in some instances, we are simply unable to determine exactly what is meant by the Hebrew (Masoretic) texts. Thus, each and every translator of an ancient text reaches a point where he or she has to make a decision according to the available information, striving to what Nida and Taber call "the closest natural equivalent." The mindset of the translator, the intended target population, and the original text that is used, are all contributing factors that result in the wide range of translations of Bibles available today. Nida and Taber, in 1982, estimated that at least three thousand people are engaged primarily in the translation of Bibles into some eight hundred languages. Therefore, today there is no fixed entity that all Christians could call "The Bible", but rather many interpretations of an array of incomplete sources. For this reason I have refrained from using the term "The Bible" in this publication, and opted for the term "Bibles" even in the title. As new information becomes available through discoveries and research, our understanding of the Masoretic texts is continually changing. We therefore could say that the true "Word of God" is being revealed to us on a continual basis. I believe that it is only at the end of time, as we know it, that it will be revealed completely.

Palestine has a wide range of habitats, resulting in a relatively wide variety of bird species for such a small area. Its location on a main migratory route means that numerous

---

3   Wikipedia contributors. "Leningrad Codex." Wikipedia, The Free Encyclopedia. Wikipedia, The Free Encyclopedia, 24 Jul. 2011. Web. 6 Aug. 2011.

migrants augment the resident birds, at least twice per annum. All of the above result in an "ornithological jig-saw puzzle" unsurpassed anywhere else in the world.

As shown in par. 2, the scientific study of the animal kingdom, and the documentation of findings, only really began in the nineteenth century. Before that time, specific names were only given to those species that were very obvious and/or of practical importance. As the authors of the Old Testament books lived and wrote thousands of years before such studies and documentation took place, they would not have been able to, or be interested in, distinguishing between birds with similar appearances. To this should be added the further complication that in many instances the bird is used figuratively, in an attempt to illustrate a point to the original readers. Which brings us to an extremely important point: When translating, the translator has to continually keep in mind that the author wrote in his own language (called the source language), and with his original audience in mind. He did not, by any stretch of the imagination, have us in mind.

6.3 An illustration

To illustrate all of the above practically, we shall investigate the dilemma of the Hebrew word *raham*. As the Hebrew word occurs only twice, namely in Lev 11:18 and Deut 14:17, which are the lists of "unclean" birds, and we therefore have no further clues in the context (other than it is indeed a bird) as to what exactly is meant, it is basically a subjective decision by the translator as to which bird is meant. Some translations e.g. JB, NIV and A33 translate the word as the Osprey, *(Pandion haliaetus)*. Koehier & Baumgartner (p. 886) prefer the Egyptian Vulture, *(Neophron percnopterus)*. I would go along with the latter as the primary diet of the Osprey would make it less abhorrent as the Egyptian/Carrion Vulture. A83 changed it from *Visvalk* meaning "Fish Hawk" or "Osprey" in the 1933 translation to *Swaan* which means "Swan" in the 1983 translation. The LB connects the Osprey to the Hebrew *oznijah*.

6.4 In a Nutshell

With all the above in mind, allow me to begin by making a (to many, maybe shocking) statement by stating that today, there exists no such thing as "The Bible." Scholars and laymen alike are using this term indiscriminately without explaining or clarifying what they mean. For example: Christian and Jewish believers would obviously not agree on what should be considered to be the "Word of God." But less obvious is the fact that my initial statement is relevant, even when only Christian believers are involved. This phe-nomenal statement is based on two sets of realities:

- Firstly: The variables that were presented during the development of the Scriptures as seen in this Chapter. The original documents have been gone for even thousands of years. Mistakes were made all along the process of copying and translating. Each and every individual involved in this process had or has his or her own personal mindset, formed according to his or her background, education, socio-economic situation, and a myriad of other factors. Some even had other agendas than pro-ducing the perfect product.
- Secondly: The differences that are present today in what one particular individual would call "My Bible", or a group would call "Our Bible." For example: Christians and Jews differ. Hundreds (if not thousands) of translations differ. Interpretations of exactly the same words differ. The dictionaries that translators are using have conflicting views on what the original authors wanted to convey to his readers. *Ad infinitum.*

So, in a nutshell, what do I believe? I believe that we are merely gazing into an ex-tremely dim image in a mirror (1Cor. 13:12), trying our best to see a clearer image of what we believe to be "The Word of God" or "The Bible." God in turn, in his infinite wisdom, is constantly revealing himself to us in greater detail. This is a process which I believe will continue until He reveals himself to us fully in the next era. In the meantime, I am thankful for those texts that I have to my disposal, and will keep on studying an as wide as possible variety, and then submit myself to Him to convince me of the truth.

In addition, I believe that God is revealing himself to mankind, not only by means of Scripture, but also by means of Nature. To me, the latter speaks simply and clearly, with-out misunderstanding and/or controversy. Even a child can look up into the night sky, or hold a sparrow and know we are dealing here with something way out of the grasp of our human minds. Therefore, to me "God" is just a word we humans use to name the incom-prehensible being that created everything from nothing. A being that allowed a myriad of species to develop, and then decided that one species should be more special than any other, by giving it what we call a "soul", and so become the crown of his/her creation.

For the past sixty years I have had the privilege of getting better acquainted with him/her. I have seen that the universe has no end, so I have come to realize that s/he has no end: Neither in time, nor in place. I have seen that what s/he created is absolutely perfect and good, so I have come to realize that s/he is perfect and good. And then I remember: the birds are displaying his/her glory *par excellence*!

As a result, I now call him/her not only "God", but also "Creator", "Omnipotent One", "All-wise One", "Father", "Mother", "Perfect One", and "Good One." But superceding all of these titles is the one I believe says it all: "Love". I see "Love" everywhere and every day. I see "Love" on the subway for example. I see it in the eyes of a grandpa, across the aisle from me, looking down at his grandson on his knee. I see it a few seats away in the comforting arm of the elderly farmer around the shoulders of his wife. Both probably on a subway for the first time in their seventy or more years, and obviously amazed by the technological wonders surrounding them. I see it in the smile of the young girl in the corner next to the door, looking up at her Romeo, who has his arms tight around her waist, both of them completely oblivious to the stares of some fellow passengers. I see "Love" whenever and wherever someone cares, and it confirms time and again: where there is "Love", there is no selfishness, no greed, no hate, no bitterness, no grudges, and no need for revenge.

WHAT IS LOVE?

"Love is when my Mommy makes coffee for my Daddy, and she takes a sip before giving it to him, to make sure the taste is OK." **Danny – age 7.**

"Love is when Mommy gives Daddy the best piece of chicken." **Elaine – age 5**.

A four year old boy whose next door neighbor was an elderly gentleman, who had recently lost his wife, saw the old man crying. The little boy went into the old gentleman's yard, climbed unto his lap, and just sat there. When his mother asked what he had said to the neighbor, the little boy said: "Nothing. I just helped him cry." **Unknown – age 4.**

BIRDS AND BIBLES
IN HISTORY

PART 3
BIBLICAL BIRDS

---

*Conducting a study of the thirty-six species
mentioned in Christian Bibles according to the seven
Orders in which they occur, by taking note
of the relevant original Hebrew and Greek texts.*

Greylag Goose *(Anser anser)*
(Photo: Tony Hisgett) [1]

CHAPTER 7
ORDER: STRUTHIONIFORMES

"The wings of the ostrich flap joyfully, but they cannot compare with the pinions and feathers of the stork. Yet when she spreads her feathers to run, she laughs at horse and rider."

Job 39: 13, 18 (NIV)²

The Emu *(Dromaius novaehollandiae)*
(Photo: Ross Berteig) ³

7.1 Flightless Birds

Gondwana, **Gondwanaland**, noun: a hypothetical landmass in the Southern Hemisphere that separated toward the end of the Paleozoic Era (roughly five hundred to two hundred and fifty million years ago) to form the present South America, Africa, Antarctica, and Australia. Gondwana included most of the landmasses in today's Southern Hemisphere, as well as the Arabian Peninsula and the Indian subcontinent.

The ratites (Latin: *ratis* which means "raft") are a diverse group of large, flightless birds (most of them now extinct), originally from Gondwanaland. Unlike other flightless birds, the ratites have no keel (as in boat) on their sternum (breastbone). Without this bone to anchor their wing muscles, they would not be able to fly even if they were to develop wings suitable for flying.

---

² THE HOLY BIBLE, NEW INTERNATIONAL VERSION®, NIV® Copyright © 1973, 1978, 1984, 2011 by Biblica, Inc.™ Used by permission. All rights reserved worldwide.

³ This image is a reproduction of the photograph taken by Ross Berteig and used with permission. Originally placed on www.flickr.com, the original photograph can be viewed at: http://www.flickr.com/photos/rberteig/4052527328/.

There are taxonomical systems that consider the various families of ratites to be orders, but the system used here uses the order *Struthioniformes* to refer to all ratites. In that case the order consists of seven families of which five are living and two are extinct. They are:

Kiwi (Apterygidae)
(Photo: Ashok Chawla) [4]

- *Struthionidae*: From Africa and Australasia. Consisting of only one living species namely the Ostrich *(Struthio camelus).* All the other species in this family/suborder are now extinct.
- *Rheidae*: From South America. Consisting of two Neotropical Rhea species. They have three toes, and their heads and necks are feathered.
- *Casuariidae*: The Cassowaries, consisting of three Australasian species that have three toes, prominent casque or helmets, and are adapted to living in rain forests.
- *Dromaiidae*: One species, the Emu *(Dromaius novaehollandiae)* from Australia.
- *Apterygidae*: The three Kiwi species from New Zealand are small to medium sized, flightless, and mainly nocturnal, with rudimentary wings, long flexible bills and no tail.
- *Dinornithidae*: New Zealand. Extinct since about 1400. The moa were eleven species of flightless birds endemic to New Zealand. The two largest species, *Dinornis robustus* and *Dinornis novaezelandiae*, reached almost four meters in height with neck outstretched, and weighed about two hundred and thirty kilograms.
- *Aepyornithidae*: The Elephant bird. It has been extinct since at least the seventeenth century. A family of flightless birds found only on the island of Madagascar. They were believed to have been over three meters tall and weighing close to four hundred kilograms. Remains of *Aepyornis* adults and eggs have been found. In some cases the eggs have a circumference of over one meter and a length up to thirty-four centimeters. The egg volume was equal to about sixty chicken eggs.

---

[4] This image is a reproduction of the photograph taken by Ashok Chawla and used with permission. Originally placed on www.flickr.com, the original photograph can be viewed at: http://www.flickr.com/photos/folley1941/4539392039/sizes/l/in/set-72157623901057170/.

CHAPTER 8
FAMILY: STRUTHIONIDAE

> "Therefore I will wail and howl, I will go stripped and naked;
> I will make a wailing like the jackals, and a mourning like the ostriches,"
> **Mi 1:8 (KJV)**

8.1 The Ostrich

The family *Struthionidae* consists of only one living species, namely the Ostrich *(Struthio camelus)*. A total of five subspecies are commonly recognized:

- *Struthio camelus australis*   Southern Ostrich, from southern Africa.
- *Struthio camelus camelus*   North African- or Red-necked Ostrich, in North Africa.
- *Struthio camelus massaicus*   Masai Ostrich, in East Africa.
- *Struthio camelus syriacus*   Arabian Ostrich or Middle Eastern Ostrich, Middle East. (This subspecies became extinct in about 1966.)
- *Struthio camelus molybdophanes*   Somali Ostrich, in Ethiopia, Kenya, and Somalia.

Of all living birds the ostrich is the tallest, (males grow up to two and a half meters tall), and heaviest (males weigh about one hundred to one hundred and thirty kilograms. An individual that weighed one hundred and fifty-seven kilograms is the highest recorded weight to date). They were formerly found all over Africa, Asia, and Southern Europe, including the Middle East. Now, unfortunately, they are endemic to Africa only. As shown above, there are five living subspecies according to their geographical distribution. The differences between them are however very slight and can be detected in their size, by their skin color, and in the size and texture of their eggs. Xenophon of Athens was a Greek historian, soldier, mercenary, and a contemporary and admirer of Socrates. He recorded the abundance of the ostrich in Assyria *(Anabasis*, i. 5). This subspecies from

Asia Minor is extinct and today all the extant ostrich races are restricted to areas in Africa. As one of the ratites they have no keel (as in ship) on their sternum (breastbone), and because of this they are not able to fly. However, they are well adapted to the plains and savanna that they inhabit. Using their long legs and neck and excellent eyesight, they are virtually ensured to detect predators and other dangers a long way off. The ostrich has the largest eyeballs of

Ostrich *(Struthio camelus)*
(Photo: Andrew Massyn) [1]

any bird alive today, with each eyeball measuring about five centimeters across. They are the only birds in which the toes have been reduced to two, namely the third and fourth.

They have no uropygial (oil) gland, thus are therefore unable to waterproof their feathers like most other birds, resulting in them getting soaked by rain or other water. Juveniles of one month and older and adults are able to run at speeds up to fifty to sixty kilometers per hour. This can be kept up for about half an hour. This makes them by far the swiftest of all cursorial birds (adapted specifically to run). Some sources state maximum speeds of about ninety seven kilometers per hour, which seems unlikely. When running at high speed, the wings are spread to assist in balancing, particularly when swerving sharply to side step a pursuer. This and other behavior patterns were observed by the author of the Old Testament book of Job.

Savanna, noun:

1. A vast plain characterized by a variety of grasses and with scattered tree growth, especially on the margins of the tropics where the rainfall is seasonal, as in eastern and southern Africa.

2. Typically a grassland region in the form of a plain with scattered trees, which often will grade into either open plain or woodland, usually in subtropical or tropical regions.

---

[1]	This image is a file from the Wikimedia Commons. Full details of the file can be found at: http://en.wikipedia.org/wiki/File:Ostriches_cape_point_cropped.jpg. Andrew Massyn, a Wikimedia Commons user, and the copyright holder of this work, released this work into the public domain. This applies worldwide. In some countries this may not be legally possible; if so: he grants anyone the right to use this work for any purpose, without any conditions, unless such conditions are required by law.

Ostriches occur in dry savannas, semi-desert and proper desert areas, avoiding grass and other vegetation higher than about one meter. When not breeding they occur in flocks of about thirty-five birds, with hundreds (as many as six hundred and eighty individuals have been counted in a flock) gathering occasionally at waterholes in deserts although they do not need to drink water. They are omnivorous, with all sorts of plant material, insects, lizards, fruits, seeds, and nuts included in their diet. They swallow pebbles and bits of sand to assist in the digestion of their food, and have a tendency to pick up small bright objects. Many a tourist visiting an ostrich farm, have been scared stiff by an ostrich suddenly pecking at their sunglasses or trinkets.

In drier years when food is scarcer, one male may have a harem of up to three females, but they are otherwise monogamous in good years. Their simple nests consist of a scraped area of about three meters in diameter in sandy soils. Clutches of up to seventy-eight eggs have been recorded, (the average being about thirteen eggs). Each female lays three to eight eggs in the communal nest. This will depend on her age and rank in the flock. The eggs are laid in the evening or late afternoon, approximately forty-eight hours apart. The eggs average one hundred and thirty by one hundred and sixty millimeters in size, weigh about one and a half kilograms, have a thick shell (about two millimeters), and are the equivalent of about twenty-four domestic chicken eggs. The eggs are the largest of any living bird in actual size. However, in relation to the size of the bird, ostrich eggs are relatively small. Measured in this way, the Kiwi from New Zealand has the largest egg of any living bird. Incubation lasts for forty to fifty days, and is done exclusively by the dominant female, the so-called "major hen", and is assisted by the male. The female incubates by day and the male at night. The reason for this is that the more cryptical (concealing) colors of the female blend in well with the desert-like surroundings and therefore the nest is better protected during the day. Temperatures are relatively high in their typical habitat, so the eggs are often left partly covered by sand during the day for the sun to keep them warm. The "major hen", if not able to cover the entire clutch, will actually move her eggs to the middle of the nest to ensure their incubation.

Newly hatched chicks are about thirty centimeters high, strongly built, with a robust disposition, and precocial (relatively mature and mobile from the moment of hatching and instinctively able to eat by themselves). Large flocks of chicks of up to three hundred or more individuals are often found. This is the result of the fact that when two family groups meet, the parents will fight for the guardianship of the chicks, and the victorious pair will then take off with and tend to both broods. The chicks take about three to four years to reach full maturity. The habit of feigning death by chicks when they sense danger, have probably led to the misleading fabrication that ostriches bury their heads in the sand at the approach of eminent danger.

8.2 Utilization by Man

Throughout the ages ostriches have been utilized by humans in some way or another. Their meat, skin, feathers and eggs have been sought after at different times for different reasons. Ostrich farming was known in ancient Egypt. Both Mesapotamian and Egyptian art show woman and royal families adorned with ostrich feathers. Due to the perfect symmetry of the feathers, the Egyptians used them as a symbol of justice. The Zulu people of South Africa use the feathers in their traditional warring attire. Because of their softness and unique structure, the white wing and tail feathers of the male became so sought after by the fashion world in the eighteenth century, and because no thought was given to sustainable utilization, the populations were virtually exterminated in the Middle East, Far North Africa, and South Africa. This eventually led to the establishment of commercial ostrich farms. The first of such modern operations was set up in 1833 in Oudshoorn, a village on the plains between the Auteniqua and Swartberge (Black Mountains) mountain ranges in the Cape Province of South Africa. Soon others followed in Algeria, Australia, and elsewhere. The industry collapsed after World War I, but about ninety thousand domesticated birds have remained in South Africa, from which high quality leather is currently produced. Other than the obvious nutritional value of the eggs, the egg shells of ostriches are being utilized in the manufacture of simple bracelets and necklaces by many African peoples. Until very recently, the nomadic Bushman people, living in the wild areas of Southern Africa, used them extensively as receptacles for carrying water with them, and also for the storing of water underground in and amongst the dunes for later use.

The biblical authors had probably known the subspecies *Struthio camelus syriacus* that inhabited the Syrian and Arabian deserts until the beginning of the twentieth century, "but after the First World War the proliferation of firearms, coupled with the availability of motorized transport, led to the devastation of populations and virtual extinction of the subspecies by 1941. The last record was of an individual found drowning in Jordan in 1966." (del Hoyo, et al, Vol. 1, p. 82).[2] Since the 1970's, chicks have been introduced into the southern Negev desert in Israel from a breeding program in the Yotvata Hai-Bar Nature Reserve.

In some countries, people race each other on the backs of ostriches. Within the United States, a tourist attraction in Jacksonville, Florida called *The Ostrich Farm* opened up in 1892 and its ostrich-drawn cart races became one of the most famous early attractions in the history of Florida. The Highgate Ostrich Show Farm is a tourist attraction in the

---

[2]   del Hoyo, J., Elliot, A. & Sargatal, J. eds. (1992). Handbook of the Birds of the World. Vol. 1. Ostrich to Ducks. Lynx Edicions, Barcelona. Used with permission.

town of Oudtshoorn in South Africa. They opened to the public in 1938 and present daily ostrich races.

8.3 Biblical References

In Biblical times ostriches were common residents in Africa and Arabia. The Assyrians used ostrich-egg cups as far back as 3000 BC. In the Roman Empire, roasted ostrich was often served at banquets and feasts, and their physicians are known to use ostrich fat as a drug. When the ostrich is mentioned in Old Testament Hebrew, the word *ya'anah* is used, with *bat ya'anah*, meaning "daughter of ostrich", occurring several times, and probably referring to the females of this family.

The date of the book Job is unknown, with 400–600 BC being a popular estimate. In Job 37–41 God is indicating his majesty to Job by using examples from nature. In Job 39:13–18 the behavior of the female ostrich (Hebrew: *renanim*), is described in detail, and it is clear that the author had observed the habits and behavior of the bird correctly. The text also mentions a stork and some avian terms, and has been interpreted in different ways. A reference to the apparent harsh treatment of the chicks by the female is found in Lam 4:3. In Is 43:20 the ostrich is pictured as praising God for an abundant season in the desert.

In Is 13:21, Is 34:13 and Jer 50:39, the ostrich is associated with the desolation and loneliness of ruins and death. The voice of the ostrich, which is described as a deep booming sound, similar to that of a lion at long distance, is used in Micah 1:8 to describe the cry of a mourner in deep distress. In Job 30:29 the author describes his loneliness by stating that the ostriches are his only friends.

Ostriches are well known to eat virtually anything. Normally their diet would include all types of plants, grass, berries, seeds, succulent plants, small reptiles, and insects. It is therefore not surprising to find the ostrich listed under the "unclean" birds in Lev 11:16 and Deut 14:15. Often small stones and bits of sand are swallowed and kept in the gizzard to assist in the digestion process. They are often attracted by shining objects, and this has lead to the erroneous belief, as mentioned by Shakespeare for example, that they can digest metal.

CHAPTER 9
ORDER: FALCONIFORMES

"But he didn't split the birds.
Vultures swooped down on the carcasses, but Abram scared them off."

Gen 15:10–11 (MSG)[1]

9.1 Diurnal Birds of Prey

Diurnal, adjective:
1. Happening during the day or daily.
2. (Of flowers) open during the day and closed at night.
3. (Of animals) active during the day.
4. Compare **Nocturnal**, adjective: Active at night. (See par. 15.1).

As stated in par. 4.7, DNA profiling has resulted in the taxonomy of birds being in a flux at the moment (July 2011). Here is a perfect example, as described in the Wikipedia article on the proposal that a new order *Accipitriformes* should be formed. The introduction to this article is quoted here:

"The *Accipitriformes* is an order that has been proposed to include most of the diurnal birds of prey: Hawks, Eagles, Vultures, and many others, about two hundred and twenty-five species in all. For a long time, the majority view has been to include them with the falcons in the *Falconiformes*, but some authorities have recognized a separate *Accipitriformes*. A recent DNA study has indicated that falcons are not closely related to the *Accipitriformes*, being instead related to parrots and passerines. Since then the split (but not the placement of the falcons next to the parrots or passerines) has been adopted by the American Ornithologists' Union's South American Classification Committee (SACC), its North American Classification Committee (NACC), and the International Ornithological Congress (IOC). The DNA-based proposal and the NACC and IOC classifications include the New World Vultures in the Accipitriformes, an approach followed in this

---

article. The SACC classifies the New World Vultures as a separate order. The placement of the New World Vultures has been unclear since the early '90s."[2]

For the purposes of this publication there is no need to get involved in the debate. More important here is to distinguish between the following:

- Those families not found in Christian Bibles, which I call "Non-Biblical" families. They are mentioned in the next paragraph, par. 9.2.
- Then those families found in Christian Bibles which I call "Biblical Families" and are mentioned in par. 9.3.
- Then the *Accipitridae* are dealt with in detail in Chapter 10, the *Falconidae* in Chapter 11, and for interest sake the *Pandionidae* in Chapter 12.

9.2 Non-Biblical Families

- *Cathartidae*: This family consists of the seven species of the so-called New World Vultures. They are found only in North and South America. Some examples are: the Turkey Vulture *(Cathartes aura)*, American Black Vulture *(Coragyps atratus)*, California Condor *(Gymnogyps californianus)*, and the Andean Condor *(Vultur gryphus)*. In appearance they are similar to the so-called Old World Vultures, found in Europe, Africa and Asia, but with two obvious differences. Firstly, they have no functional hind toes, and secondly, because their nostrils are not separated internally, it is possible to look through the scull of the bird

California Condor *(Gymnogyps californianus)*
(Photo: Dr. Marcel Holyoak) [3]

2   Wikipedia contributors. "Accipitriformes." Wikipedia, The Free Encyclopedia. Wikipedia, The Free Encyclopedia, 13 Jul. 2011. Web. 6 Aug. 2011.

3   This image is a reproduction of the photograph taken by Dr. Marcel Holyoak and used with permission. Originally placed on www.flickr.com, the original photograph can be viewed at: http://www.flickr.com/photos/maholyoak/5851473024/sizes/l/in/photostream/.

from one side to the other. As most Bibles originated in the Middle East, which is part of the Old World (Europe, Africa, and Asia), we shall only be discussing the vultures found in the latter region.

Secretary Bird *(Sagittarius serpentarius)* *(Photo: Kevin Hughes)* [4]

- *Sagittariidae*: the Secretary Bird *(Sagittarius serpentarius),* is the only species in this family. As it endemic (occurs only in a particular place) to sub-Saharan Africa, it will not be dealt with here.
- *Pandionidae*: the Osprey *(Pandion haliaetus)* is the only species in this family. Four sub-species have been identified. The distinctions are based mainly on the geographical regions in which they are found globally. The diet of ospreys consists almost exclusively of live fish, for which its body has undergone some special adaptations. This family is dealt with in Ch.12.

9.3 Biblical Families

- *Accipitridae*: This is the most heterogeneous and largest family in this order, with about fifteen subfamilies including:
    - "True" or "Booted" Eagles (twenty-nine species),
    - Hawks (forty-nine species),
    - Buzzards (twenty-one species),
    - Kites (twenty-one species),
    - Cuckoo-hawks (two species),
    - Bazas (three species),
    - Sparrow-hawks (twenty-one species),
    - Goshawks (twenty-two species),
    - Chanting-goshawks (three species),

---

4   This image is a reproduction of the photograph taken by Kevin Hughes and used with permission. Originally placed on www.flickr.com, the photograph can be viewed at: http://www.flickr.com/photos/48881750@N03/4497142300/.

- Harriers (eight species),
- Harrier-hawks (two species),
- Marsh harriers (five species),
- Old World Vultures (fifteen species, including four Griffon species),
- Fish-eagles (five species),
- Sea-eagles (four species),
- Snake-eagles (five species),
- Serpent-eagles (eight species),
- Hawk-eagles (thirteen species).

All together a total of two hundred and thirty-seven species.

They are found in a wide range of body sizes: from small, for example: The African Little Sparrow-hawk *(Accipiter minullus)* which weighs a mere seventy-five to eighty-five grams and is twenty-three to twenty-seven centimeters in length, to very large, for example: the Condors *(Vultur* and *Gymnogyps Spp.)*, and the Griffon Vultures *(Gyps Spp.)*, measuring up to one and a half meters, weighing up to twelve and a half kilograms, with wingspans up to three meters in the case of the Himalayan Griffon *(Gyps himalayensis)*. This family is dealt with in Chapter 10.

- *Falconidae:* This family includes the Falcons, Falconets, Kestrels, Forest-falcons, and Caracaras, for a total of about sixty species worldwide. This family is dealt with in more detail in Ch. 11.

The Hebrew word *'ajit* occurs six times in the Old Testament. In all instances it is used collectively for all the diurnal birds of prey. In Gen. 15:11, Job. 28:7 and Is. 46:11 the word is used in this manner, but the contexts do not provide us with any further insights in determining specific species. In Is. 18:6 they are mentioned twice and it suggests that these birds are associated with mountains. The author has observed them building their nests on the cliffs in summer, and here implies that grapevine cuttings are used as nest building material.

Ezk 39:4 literary mentions "birds of prey of birds", usually translated as "all the birds of prey." The context confirms that birds of prey are mainly carnivorous.

In Jer 12:9 *'ajit* is utilized twice. The second occurrence is used conventionally. In the first instance the Hebrew literary states "the birds of prey Hyena." For Hyena, the Hebrew word *tsabu'a* is used, which is derived from the verb *tsabah*, meaning "to dip or dye." As a result it is reasoned that the word refers to the Spotted Hyena *(Hyaena hyaena)*. Therefore, translators have opted to translate *'ajit tsabu'a* with "spotted birds of prey." This would seem to be a reasonable conclusion for two reasons:

- Firstly, many raptors, as found in Palestine too, are spotted, if not as adults, e.g. Common Kestrel *(Falco tinnunculus)*, Short-toed Eagle *(Circaetus gallicus)*, and several Harrier, Sparrowhawk, and Buzzard species, then at least in their juvenile and sub-adult stages, e.g.: Black Kite *(Milvus migrans)*, Golden Eagle *(Aquila chrysaetos)*, Spotted Eagle *(Aquila clanga)*, Osprey *(Pandion haliaetus)*, and Egyptian Vulture *(Neophron percnopterus)*.
- Secondly, when reading the Old Testament, it becomes quite clear that many of the authors were excellent observers of nature, and would, without difficulty, notice the similarity between the colors of a hyena and those of these birds of prey.

However, we should always keep in mind that Bibles are not scientific publications, and that the authors had a specific message in mind. To convey their message to their readers, they, however, made excellent use of the objects with which they and their readers were familiar. It is therefore not surprising to find them using a collective term like "Birds of Prey", as the authors themselves were not professional Ornithologists, but could rather be described as amateurs using the knowledge available at the time to reach their objectives. The messages in the particular contexts would always remain paramount to them.

With the basic hypothesis of this book in mind, it should be noted that Jeremiah worked before the fall of Jerusalem to the Babylonians in 587 BC, two hundred years before the birth of Aristotle.

CHAPTER 10
FAMILY: ACCIPITRIDAE

> "Yes, God will rise up a faraway nation against you,
> swooping down on you like an eagle."
>
> **Deut 28:49 (MSG)**[1]

10.1 Eagles

Aerie, noun:

1. The nest of a bird of prey, such as an eagle or a hawk.
2. A lofty nest of any large bird.
3. A house, fortress, or the like, located high on a hill or mountain.

"*Aerie*" is the fifth album by American singer-songwriter John Denver. It debuted on the *Billboard 200* album charts on 4 December, 1971, hitting number 75. It includes his own composition *The Eagle and the Hawk.* (See Introduction, par. e).

To the layman, eagles are best known for their impressively large size, their strength in killing, and their speed. Some of their most distinguishing features are:

- They have short and strongly hooked bills.
- Their eyes are relatively large, slightly larger than human eyes, but they can see up to three and a half time better than a human with perfect vision. For example: an eagle can see a moving rabbit from two kilometers away, and it can see a fellow eagle soaring for up to eighty kilometers away.
- They have short strong legs that are covered in feathers up to their toes.
- They have powerful feet with opposable hind toes.
- They have long sharply curved claws which are deadly efficient when grasping live prey.

---

[1] Scripture taken from The Message. Copyright © 1993, 1994, 1995, 1996, 2000, 2001, 2002. Used by permission of NavPress Publishing Group.

- Their well developed wings make them powerful fliers, and they are well known for their maneuverability and speed. For example: slow-motion footage of an eagle pursuing prey in a densely wooded area makes for some fascinating viewing.
- Many of the larger species have broad and long wings which make them superb soarers.

Major new research into eagle taxonomy suggests that the important genera *Aquila* and *Hieraaetus* are not composed of nearest relatives, and it is likely that a reclassification of these genera will soon take place, with some species being moved to *Lophaetus* or *Ictinaetus*

Carnivore: noun, meaning "meat eater" (Latin: *carne* means "flesh", and *vorare* means "to devour"), is an organism that derives its energy and nutritianal requirements from a diet consisting mainly or exclusively of animal tissue, whether through predation or scavenging.[2]

Eagles are mainly carnivorous, catching live prey, or scavenging on carrion (the carcass of a dead animal). Some groups have specialized diets, for example:

- Snakes: the Old World Snake-eagles *(Circaetus),* and the New World Serpent-eagles *(Spilornis).*
- Birds: Sparrowhawks and Goshawks.
- Bats: the Bat Hawk *(Macheiramphus alcinus),* swallowing its prey whole in flight after hunting in tandem with his/her mate.
- Fish: Fish-eagles *(Haliaeetus)* being piscivorous (a carnivorous animal which primarily eats fish).
- Mammals, rodents and reptiles of all sizes. Anything up to the weight of the bird will be targeted, including rabbits, hares, adult monkeys, goats, lambs, etc.

Some species have undergone special adaptations according to their diets. For example, the tarsal joints of the African Harrier-hawk or Gymnogene *(Polyboroides typus),* is able to bend seventy percent backwards and thirty degrees sideways. In addition to this, it has a relatively small and narrow head, all of which allows it to reach deep into holes and crevices, taking eggs and catching nestlings, bats, squirrels, etc. The feet and talons (claws) of eagles are mainly used to kill the prey by stabbing the vital organs, but pecks to the head or neck may kill larger vertebrates. Smaller prey is either swallowed whole or is first torn to pieces and devoured piece meal.

---

2   Wikipedia contributors. "Carnivore." Wikipedia, The Free Encyclopedia. Wikipedia, The Free Encyclopedia, 2 Aug. 2011. Web. 6 Aug. 2011.

Eagles lay comparatively small clutches of eggs (usually one or two), and have long incubation periods (forty to sixty days). With most species the second chick is killed by the first (the so-called "Cain and Abel" behavior). With other species, only the first chick is fed, leaving the second one to die of starvation. The chicks are altricial (requiring nourishment and care), staying in the nest for anything from seven to seventeen weeks. In addition, they have long post-nestling dependence periods (from about six weeks up to as long as eleven months). All of this resulting in eagles having relatively slow reproductive rates for a bird.

The figure of the eagle has through the ages, and to this day, been a popular ensign in both the military and civil services. Eagles were employed as symbols of power in old Chinese dynasties. Raptor representations are common on bas-reliefs (a method of sculpting which entails carving or etching away the surface of a flat piece of stone or metal) in the temples and tombs of the ancient Egyptian civilization. The Ptolemaic rulers of Egypt used it as their seal. Cyrus, founder of the Persian Empire, was closely associated with the eagle, as Is 46:11 clearly illustrates. The eagle was employed in similar fashion by the Assyrians and Romans. The Aquila was the eagle standard of a Roman legion. Napoleon I used the Roman Golden Eagle as the symbol of his new French empire.

Eagles serve as the national emblems of countries like Poland, the German Reich, and the United States of America, who have chosen the Bald Eagle *(Haliaeetus leucocephalus)*, as their national bird. No less than thirty countries have included one or more eagles in their coat of arms.

The Hebrew word *nesher* is often translated as "eagle", but this is as much a generic word as the English. As many of the references are used figuratively, it does not give many clues as to the specific species, and could include any large raptor. Dictionaries therefore mention that it could without difficulty be translated as "vulture." Micah 1:16 is a case in point. It was customary for mothers, who had received news of the death of a child to, in a fit of grief, pluck out or shear off their hair until bald. The author had seen the bald heads of the vultures, and uses it to emphasize the extent of the grief. Modern Hebrew has followed this line of though, and calls the Griffon Vulture *(Gyps fulvus)* by this name.

We unfortunately cannot know which specific eagle(s) are referred to in the Bibles. In the contemporary Middle East there are resident species like the Golden Eagle *(Aquila chrysaetos)* and Bonelli's Eagle *(Aquila fasciata),* and many migrant species like the Steppe Eagle *(Aquila nipalensis),* Spotted Eagle *(Aquila clanga),* and so on.

10.1.1 Biblical References

In both the lists of "unclean" birds found in Lev. 11:13–19 and Deut 14:12–18, *nesher*, is the first to be mentioned. The "clean birds" are not mentioned in detail. As many of the "unclean birds" are birds of prey, it would seem that a bird would be classified as such purely on account of its diet. Carnivorous birds would not be suitable as food or sacrificial purposes, and those scavenging on carrion would be especially obnoxious.

In eight of the twenty-five references to the Eagle in the Old Testament, namely: Deut 28:49, 2 Sam 1:23, Job 9:26, Jer 4:13, Jer 48:40, Jer 49:22, Lam 4:19 and Hab 1:8, the authors are using the exceptional speed with which the eagle swoops down on its prey to illustrate a particular point in their writings.

Bonelli's Eagle *(Aquila fasciata)*
(Photo: Paul Asman and Jill Lenoble) [3]

In Ex 19:4 the eagle is portrayed as a source of protection. Job 39:27, Jer 49:16 and Obd 1:4 state that eagles build their nests high.

In Ezk 1:10 and 10:14 the author merely mentions the face of the eagle, but in Ezk 17:3 and 17:7, Dan 4:30 and Dan 7:4 correctly describes a number of outstanding features of eagles. More recent translations, e.g. NIV and TEV, mention an eagle further on in Ezk 17:9. It seems that this translation has been decided upon as a result of the context created in 17:7. However, the Hebrew text itself does not mention specifically who will be executing the action.

Ps 103:5 suggest that eagles have high levels of vitality; Prv 23:5 and Is 40:31 suggest that they are capable of reaching high altitudes; and Prv 30:19 that their flying skills remain a mystery to the author.

As mentioned above, eagle chicks have long post-nestling dependence periods, but it is well known that the chicks exercise their wings extensively while still in the nest,

---

3   This image is a reproduction of the photograph taken by Paul Asman and Jill Lenoble on the Golan Hights and used with permission. Originally placed on www.flickr.com, the original photograph can be viewed at: http://www.flickr.com/photos/pauljill/3497007871/

and are quite capable of flying when leaving the nest. The description in Deut 32:11 can therefore not be taken literally.

In the New Testament the eagle is mentioned explicitly in Rev 4:7 and 8:13. In Rev 12:14 the wings of an eagle are mentioned.

10.2 Vultures

To the layman in Biblical times, the vultures were probably the most abhorrent of all the birds. Even to this day they are more often than not associated with the stench and filth of decaying carrion (dead and rotting flesh). Their value as scavengers in preventing the spread of disease, like Anhtrax, especially in undeveloped areas, was not appreciated then, and even to a large extent, today. Education is the key element in environmental protection. In South Africa, Dr. Gerhard Verdoorn is a Board member of BirdLife International, BirdLife South Africa, and the Endangered Wildlife Trust. He has devoted most of his adult life in educating the public in general on the virtues of vultures, and through his Griffon Poison Information Centre, more specifically farmers and pesticide users. He has been successful in many projects like being instrumental in changing the perceptions of farmers in the Northern Cape Province to such an extent that many raptor species have returned to the area.

A characteristic for which vultures are well known is their almost bare heads and necks, which is covered with short, sparse, downy-like feathers only. This was probably acquired during their evolution as a result of their carrion-feeding habit of reaching deep into carcasses. Their necks appear long, and can be withdrawn almost completely into the large collar surrounding it. Their powerful bills are longer that those of the Eagles, but, typical of a bird of prey, strongly hooked. The Palm-nut Vulture *(Gypohlerax angolensis)* being an interesting exception, in that its diet consists largely of the fleshy pericarp of the Oil Palm *(Elaeis guineensis)* and *Raphia* palm's fruit, other fruits, grain, and even crabs and fish. Vultures can be divided into three groups according to their size:

- Exceptionally large species, measuring up to more than one meter in length, and weighing up to eleven kilograms, placing them among the largest of all flying birds. They usually have comparatively dark, drab plumages, and are renowned for their soaring at tremendous heights (occasionally spotted by commercial airline pilots cruising at ten thousand meters), their exceptional eyesight, and the great distances covered on a daily basis. They are gregarious (sociable), nesting colonially, and usually on very high mountain cliff edges. They congregate at mammal carcasses in large numbers, feeding on the intestines, viscera (organs in the cavities of the body, especially those in the abdominal cavity) and soft muscles. In Africa, game rangers for example use the sight of circling vultures as a tell-tale sign to locate carcasses caused by poaching. These species include the Condors *(Vultur* and *Gymnogyps Spp.)*, and the Griffon Vultures *(Gyps Spp.)*. The California Condor *(Gymnogyps californianus)* reached the conservation and ornithological headlines world wide when the entire wild population of twenty-one birds were captured in 1987, and successful captive breeding programs resulted in the captive population reaching fifty-two in 1991. The first two birds were released into the wild the following year.

- The medium to large species, e.g. the King Vulture *(Sarcoramphus papa)* and Lappet-faced Vulture *(Torgos tracheliotus),* tend to have more brightly colored plumages and shorter necks. They have smaller ranges when foraging, and are not found in large numbers at carcasses. They specialize in feeding on the skin and tough tissues e.g. sinews and tendons, of the carcass.
- The smaller species e.g. the Turkey Vulture *(Cathartes aura)* and Egyptian Vulture *(Neophron percnopterus)* have more tolerant diets, which make them less dependent on large carcasses. Should they visit a carcass, they feed on leftovers after the larger vultures have left.

Incubation of the single or two eggs is performed by both parents and takes seven to eight weeks. The chicks are altricial (helpless when hatched and requiring parental care), staying in the nest for about twenty weeks. It is usual with other birds of prey to carry food to the nest in their feet. Vultures, however, feed their chicks by regurgitating partially digested meals initially, and later, undigested food into the nest. The chicks are fed by both parents. After leaving the nest, chicks remain dependent on the parents for about another four months. They reach full maturity at three to six years of age.

Today the Griffon Vulture *(Gyps fulvus)* is found in Spain, Africa, Eastern Europe, the Middle East, and further eastwards as far as India. They live mostly on carrion. Although they are large in size and possess great power, they seldom kill, and are even reluctant to touch prey showing any signs of life. They are usually seen soaring at great heights, scanning the area for carrion. In India they are seen around a Hindu burning-ghat (the level area at the top of a flight of steps leading to a river, used to burn corpses), waiting to feed on corpses not completely consumed by the fire.

Condors are very special to many South-American nations. They are a symbol of national identity to the people of the Andes. Therefore, they appear in the coat of arms of modern-day Colombia, Bolivia, Ecuador and Chile. Accurate drawings of the Egyptian Vulture *(Neophron percnopterus)* have been found on the walls of Egyptian tombs. Its vernacular name, "Pharaoh's chicken" would seem to imply a close relationship between the

Egyptian state and the bird. It is mentioned as an example of the use of a tool by animals, as it is known to break the shells of Ostrich *(Struthio)* eggs by throwing stones onto them.

10.2.1 Biblical References

For the second bird mentioned in the lists of "unclean" birds in Lev 11:13 and Deut 14:12, the Hebrew word *peres* is used. Hebrew dictionaries list *peres* as *Gypaetus barbatus,* and call it the "lamb vulture", derived from the earlier German Lammergeier/Lammergeyer, literally meaning "lambs vulture."

Today this species is known as the Bearded Vulture, German: *Bartgeier,* from the prominent patch of black bristles below the bill, It is occasionally referred to as the "bone breaker", from its habit of dropping larger bones from heights up to one hundred and fifty meters onto flat rocky ossuaries, breaking them into pieces small enough to swallow. The verb in Biblical Hebrew from which this word stems means: "to break bread." Modern Hebrew is consistent with this interpretation. A number of translations e.g. NIV, JB and LB simply use the more generic "vulture."

The third "unclean" bird is called *ozniyat* in the Old Testament Hebrew. In Modern Hebrew we find *ozniyat hanegev (Torgos tracheliotus negevensis)* usually translated as the Lappet-faced Vulture, and *ozniyat shekhora (Aegypius monachus)* usually translated as the Black Vulture. The latter should however not be confused with the American Black Vulture *(Coragyps atratus).* Old Testament Hebrew dictionaries agree with this interpretation. With all the above in mind, a meaningful translation of Lev 11:13c

American Black Vulture *(Coragyps atratus)*
(Photo: Rus Koorts) [4]

---

[4] This image is a reproduction of the photograph taken by Rus Koorts from Pretoria, South-Africa and used with permission.. Originally placed on www.flickr.com, the original photograph can be viewed at: http://www.flickr.com/photos/ruslou/408654164/sizes/o/in/photostream/

and Deut 14:12b would be as follows: "the Eagle, the Bearded Vulture, and the Lappet-faced Vulture."

In Prv 30:17 the context determines that the Hebrew word *nesher* should rather be translated with "vulture." In Lev 11:18 we find the Hebrew word *raham*, which is referred to by some dictionaries as *(Vultur percnopterus)*, which is more probably meant to be the Egyptian Vulture *(Neophron percnopterus)*.

From the above it is clear that, because of a lack of evidence from the context in which a bird is mentioned, we are, more often than not, unable to connect specific Hebrew words accurately with one particular bird species.

In the New Testament the contexts of both Mat 24:28 and Luk 17:37 suggest that vultures are the birds involved, but no further clues as to the specific species are given.

10.3 Kites

GISS/JIZZ: noun. Acronym donating "General Impression of Size and Shape" (pronounced jizz), is a term used by birders when making notes about a bird's appearance. The "GISS" or "JIZZ" of a bird can make it easy to narrow down the species for identification purposes. The acronym GISS was first used by the military in WW II. It was found that anti-aircraft gunners often fired on friendly fighter planes. There arose a need to distinguish between friendly and enemy planes within a split second to aid them in the decision to fire or not. They were trained to do this by recognizing the GISS of different planes as a first step in more accurate identification. Birders have adopted this strategy by getting a first impression of the physical size, shape and proportions of a bird, which will point them to the family or group in which the particular species can be found.

The long narrow angled wings and distinctive forked tails of the kites give them a distinctive GISS that make them easy to identify in flight. Kites are medium-sized but lightly built birds of prey, with relatively small heads, partly bare faces, short beaks, and weak legs. Kites occur worldwide in warm regions. Most kites feed mainly on carrion but some live on insects while others are primarily scavengers. They are known to also eat rodents and reptiles, and a few are strictly snaileaters. Kites are buoyant in flight, spend a great deal of time soaring, slowly flapping and gliding with wings angled backwards. Several species are as graceful as Terns *(Sternidae)* because of their hovering skills.

Three kite species are known to the Middle Eastern region. The Black Kite *(Milvus migrans),* had unfortunately been exterminated in Israel in the 1950's due to poisoning by the extensive use of agricultural pesticides in the drive to develop the wild areas by the kibbutzim system (collective communities in Israel that was based on agriculture). In the early 1970's the Black Kite recolonized as a breeder, but can be seen particularly as a winter visitor. They gather in large migratory flocks to cross the sea straits. A total of almost thirty-seven thousand birds were recorded at Eilat in the spring of 1980. They are migratory, especially the subspecies *Milvus migrans migrans,* which winters in sub-Saharan Africa as far south as South-Africa, and to a lesser extent in the Middle East. They are opportunistic hunters and are more likely to scavenge. The number of Red Kites *(Milvus milvus)* had been reduced to such an extent that only the occasional vagrant is seen in the Middle East, Finland, and Libya. The Black-shouldered Kite or Black-winged Kite *(Elanus caeruleus)* should not be confused with the Australian Black-shouldered Kite *(Elanus axillaris)*. It is known to have adapted well to agricultural areas, bush clearings, and heavily grazed areas, but still remains rare to the Middle East. Kites do hunt from a perch, but are widely known for their hovering skills when hunting.

Black Kite *(Milvus migrans)*
(Photo: Andreas Trepte) [5]

10.3.1 Biblical References

Lev 11:14 and Deut 14:13 mentions a bird by the Hebrew name of *daʾah*, which is though to be the Red Kite *(Milvus milvus milvus),* a sub-species found in Central Europe and around the Mediterranean. The stem of the Hebrew verb consisting of the same

---

[5]   This image is a file from the Wikimedia Commons. Full details of the file can be found at: http://en.wikipedia.org/wiki/File:Milvus_migrans_front(ThKraft).jpg. I, Endreas Trepte, alias ThKraft, a Wikimedia Commons user, am the copyright holder of this work, hereby publish it under the following license: the Creative Commons Attribution-Share Alike 2.5 Generic license, on 31 March 2008.

consonants, describes the action of a wolf sneaking up to its prey, or a hovering raptor pouncing on its prey. Lev 11:14 and Deut 14:13 also mentions a bird by the Hebrew name of *'ajjah*, which is thought to be the Black Kite *(Milvus migrans),* as the name might be an imitation of its cry. There are seven sub-species of which two are found in the Middle East. *Milvus migrans migrans* migrates to sub-Saharan Africa, and *Milvus migrans aegyptius* is found mainly in Egypt.

In Job 15:23 a word is found that is very close to that of the Black Kite, but it cannot be safely translated as such. The Black Kite is mentioned in Job 28:7, associating it with exceptional eyesight.

CHAPTER 11
FAMILY: FALCONIDAE

> "Does the hawk take flight by your wisdom and
> spread its wings toward the south?"
>
> **Job 39:26 (NIV)**[1]

11.1 Falcons

This is an extremely diverse family. It has been divided into two subfamilies namely:

- *Falconinae*, comprising the Falcons (twenty-four species), Falconets (eight species), and Kestrels (thirteen species).
- *Polyborinae*, consisting of the New World Caracaras and the highly secretive Forest-falcons, (sixteen species).

To the layman, falcons are known for the following characteristics:

- Relatively small and light bodies.
- Long pointed wings.
- Short necks.
- Strong hooked bills, with tomial teeth.
- Sharp curved talons.
- Exceptional eyesight and avian skills.

The Peregrine Falcon *(Falco peregrinus),* is considered by some to be the fastest bird, reaching speeds up to one hundred and eighty kilometers per hour in a stoop. However, this is debatable, as the Common Swift *(Apus apus)* is said to be capable of speeds up to two hundred and sixteen kilometers per hour for short bursts. The tomial teeth (a sharp

---

triangular-shaped downward pointing projection found at the outer edge of the upper mandible, also known as a "nook") of the true falcons are not a well-known characteristic. Their legs and talons differ depending on the primary prey of the particular species or the group in general. Their legs are short and thick for catching relatively large prey. The talons (claws) on the hind toe and elongated middle toe are massive and efficient killers.

"Except for the high Arctic and Antarctic, falcons are found throughout the world, and although they have adapted to most habitats they are usually associated with open and Savanna-like country. They have wide distribution areas, e.g. the Common Kestrel *(Falco tinnunculus)* in Eurasia and Africa, and the Australian Kestrel *(Falco cenchroides)* on that conti-

Peregrine Falcon *(Falco peregrinus)*
(Photo: Pete Walkden) [2]

nent. The Peregrine Falcon may have the widest breeding distribution of any bird, including all continents. There are interesting exceptions. For example, the Taita Falcon *(Falco fasciinucha)* and Eleonora's Falcon *(Falco eleonorae)* who are extremely restricted in distribution and numbers. The former is found in East and Central Africa, but there are a few nesting sites in Zimbabwe (Zambezi River) and Northeastern South Africa. The latter is found in the Mediterranean, including the Middle East. The Mauritius Kestrel *(Falco punctatus)* was on the brink of extinction, with only two pairs remaining in the wild in 1974. Captive propagation since 1984 with innovative harvesting and management, including supplementary feeding, provision of artificial nest boxes, predator control, and reintroduction to secondary habitats, raised the population in the wild to at least fifty breeding pairs and over two hundred birds by 1993." (del Hoyo, et al, Vol. 1, p. 260).[3]

---

[2]   This image is a reproduction of the photograph taken by Pete Walkden and used with permission. Originally placed on www.flickr.com, the original photograph can be viewed at: http://www.flickr.com/photos/petewalkden/4607079255/.

[3]   del Hoyo, J., Elliot, A. & Sargatal, J. eds. (1992). Handbook of the Birds of the World. Vol. 1. Ostrich to Ducks. Lynx Edicions, Barcelona. Used with permission.

Typically, falcons are solitary predators, highly effective in catching birds on the wing (see par. 11.2). However, there are interesting exceptions, e.g. the Lesser Kestrels *(Falco naumanni)* which specializes in insect swarms, and hunt, nest and roost together all year around. When migrating, they form spectacular flocks of up to seventy thousand birds at a single roosting spot. Kestrels regularly hover into the wind, and are mainly insect eaters, responsible for up to ninety percent of their diet. The falconets, on the other hand, are mainly insectivorous, but exceptions do occur. For example, The African Pygmy Falcon *(Polihierax semitorquatus)* who feeds on lizards and birds like the Sociable Weaver *(Philetairus socius),* with which it is closely associated for breeding. The, often massive, weaver nests are a common sight in the Kalahari region of Southern Africa, and the Pygmy Falcon actually nests in these structures. The weaver weighs about thirty-five to forty-five percent of the falcons' weight.

The Lanner Falcon *(Falco biarmicus),* often nests on pyramids and might be the falcon on which the representation of Horus, the Egyptian sky god, or god of the heavens, was based. The desert race of the Peregrine Falcon *(Falco peregrinus pelegrinoides),* also known as the Barbary Falcon, may be a good candidate too.

11.2 Falconry

Falconry, noun: the sport of causing falcons and other birds of prey to return from flight to their trainer and to hunt some form of quarry under his or her direction.

There are two traditional terms used to describe a person involved in falconry: a falconer flies a falcon; an austringer flies a hawk *(Accipiter* and similar species) or an eagle *(Aquila* or similar species). Falconry is an ancient sport, and demands a deep understanding of the falcon's behavior and the habits of the quarry. Throughout history falconers inadvertently discovered and subsequently exploited an important principle of animal behavior: namely that animals in general, and in this case birds, often react only to a narrow band of features, in this case the lure, in their surroundings. The earliest recorded images of falconers are from Arabia and the Middle East, dating back to about 720 BC. It is estimated that it was practiced as far back as 2000 BC in places as far apart as China, India, Arabia, and Persia. The sport probably reached its peak in the sixth to seventeenth centuries in Europe and Asia. Frederick II (See par. 1.6.2) was known to be so passionate about the sport, that he abandoned a siege to fly his birds. He wrote a practical manual on the sport, setting out in detail how birds ought to be trained to the peak of perfection. The sport is very much a part of the Arabian culture, with as many as three thousand falcons being trained in the Middle East in any given year. Ninety percent of these birds are female Saker Falcons *(Falco cherrug),* but Lanner Falcons *(Falco biarmicus)* are a popular choice too. For further reading refer to the extensive article by Wikipedia entitled: "Falconry".[4]

11.3 Biblical References

According to Koehler & Baumgartner (p. 628), the Hebrew word *nets* refers to the desert subspecies of the Peregrine Falcon *(Falco peregrinus pelegrinoides),* which is common in the Middle East and found as far as Iraq and even into Iran. The word occurs only three times in the Old Testament namely in Lev 11:16, Deut 14:15, and Job 39:26. None of these contexts give us any further details, except that Job mentions that the bird is heading south, which might be a reference to a migratory direction.

The author with a falconry friend
(Photo: Tian Hattingh) [5]

---

[4] Wikipedia contributors. "Falconry." Wikipedia, The Free Encyclopedia. Wikipedia, The Free Encyclopedia, 6 Aug. 2011. Web. 6 Aug. 2011.

[5] This photograph is from my own collection.

CHAPTER 12
FAMILY: PANDIONIDAE

> "… pelican, osprey, cormorant, …"
> **Deut 14:18 (MSG)**[1]

12.1 The Osprey

Although the Osprey *(Pandion haliaetus),* is not mentioned specifically in most Bibles, it is so well known worldwide and common in the Middle East, that I could not resist including this very special bird. Del Hoyo, et al, (Vol. 2, p. 49) states: "Few other birds have invoked man's admiration and concern more than Ospreys."[2]

A piscivore is a carnivorous (flesh-eating) animal which primarily eats fish. The Osprey is essentially a hawk that has developed into a fishing specialist, and therefore laymen through the ages often refer to it as a "Fish Hawk." It should not be confused with the Fish-eagles *(Haliaeetus Spp.).* As a result of its specialized diet, the Osprey has undergone several distinctive adaptations resulting in a formidable and very specialized hunter. They for example have:

- Relatively long legs, for reaching surface-dwelling fish in the water.
- Reversible outer toes.
- Spiny foot pads to aid them in gripping and carrying fish.
- A long small intestine, aiding it in digesting fish.
- A relatively dense, compact, oily plumage for better maneuverability when wet.
- Nasal valves essential for diving.

---

[1] Scripture taken from The Message. Copyright© 1993, 1994, 1995, 1996, 2000, 2001, 2002. Used by permission of NavPress Publishing Group

[2] del Hoyo, J., Elliot, A. & Sargatal, J. eds. (1994). Handbook of the Birds of the World. Vol. 2. New World Vultures to Guineafowl. Lynx Edicions, Barcelona. Used with permission.

Although the Osprey can be found in a wide variety of coastal and marine habitats, they prefer the more shallow waters alongside open water like dams, lakes, reservoirs, marshes, and seashores. They especially prefer places that also provide safe nesting sites. In the present day Israel, the commercial fishponds, found at a large number of kibbutzim, are ideal in both respects, resulting in them being a common resident.

Ospreys usually hunt alone. They actively search for prey by flying and hovering from as low as five to as high as forty meters above the water. When a fish is spotted in a suitable position, they swoop down with raised wings and heads close to the feet at speeds up to sixty kilometers per hour. They swing their feet forward just before striking and then snatch the fish in their talons (claws). In the process, they would often submerge themselves completely in the water for up to one second. Should the prey prove to be exceptionally large, they would pause for a few seconds in the water with their wings outstretched on the water surface. During this time they are probably securing their grip on the fish. They then take off by reaching high with long and fast wing strokes, and carry the prey, head first, to a suitable feeding perch. Unlike the Kingfishers (order *Coraciiformes*) they do not kill the fish but would take their time and preen and oil themselves while waiting for the fish to die. They would then start eating the fish from the head first. The size of fish caught range from one hundred and fifty grams to twice that size, with the occasional catch weighing up to three kilograms. Their strike rate varies from forty to seventy percent. This depends mainly on the agility of the species of fish that is being hunted.

Ospreys are generally monogamous. Large nests are usually built

Osprey *(Pandion haliaetus)*
(Photo: Pete Walkden) [3]

---

[3] This image is a reproduction of the photograph taken by Pete Walkden and used with permission. Originally placed on www.flickr.com, the original photograph can be viewed at: http://www.flickr.com/photos/petewalkden/4731236534/.

from sticks, twigs, driftwood and an assortment of flotsam and jetsam. They like dead trees, but man-made structures are also utilized. They even nest on the ground, especially on islands that are safe from predators. Courtship feeding takes place, with pairs averaging about sixty copulations per clutch. Clutches consist of two to four eggs. Incubation lasts about forty days, which is performed mostly by the female. Nestling (caring for chicks in a nest) takes seven to eight weeks, with the female then remaining on the nest, and food being shared sequentially, but not equally, first by the male and then by the female and her chicks. The post-nestling dependence periods lasts for two to eight weeks. The mortality rate during this time is as high as seventy percent due mainly to hazardous migratory flights. Those birds nesting in regions that experience regular winter freezing migrate to warmer latitudes, usually further south than the local residents.

12.2 Ospreys in Britain

Because of the prominence given to the successful return of the Osprey to Britain, I have included below extracts from the Wikipedia article "Ospreys in Britain".[4] The success story of the Ospreys is often featured in the press, and these days, webcams allow enthusiasts to follow the activities at the nesting sites live. Birds fitted with satellite transmitters can be followed on-line.

"The Osprey became extinct as a breeding bird in England in 1840 and in the British Isles in 1916 but recolonized in 1954. It is generally considered that the species was absent from Scotland from 1916 to 1954. In 1954 Scandinavian birds re-colonized Scotland naturally and a pair has nested successfully almost every year since 1959 at Loch Garten Osprey Centre, Abernethy Forest Reserve, in the Scottish Highlands."

"Because of the slow geographical spread of breeding Ospreys within Scotland, in 1996 English Nature and Scottish Natural Heritage licensed a project to re-introduce the Osprey to Central England. Over six years, chicks from Scottish nests were moved to the Nature Reserve at Rutland Water in the English Midlands, where they were released. In 1999 some of the translocated birds returned after their migration from Africa and in 2001 the first pair bred. The same year, a pair from the Scottish population bred for the first time in the Lake District at Bassenthwaite Lake. These two nests were the first recorded in England for 160 years."

"Since 2001, pairs of Ospreys have continued to breed successfully at both English sites and, by 2007, nineteen Osprey chicks had fledged from the Rutland colony. Another

---

4   Wikipedia contributors. "Ospreys in Britain." Wikipedia, The Free Encyclopedia. Wikipedia, The Free Encyclopedia, 30 Jul. 2011. Web. 6 Aug. 2011.

unexpected result of the translocation project was the establishment of two nests in Wales in 2004. One was at an undisclosed location and the other at the RSPB Glaslyn Osprey Project at Pont Croesor, near Porthmadog in North Wales. In both cases the adult male, although originally from Scotland, had been translocated to Rutland. Another significant milestone occurred in 2007 when a female Osprey, having fledged from one Rutland nest three years previously, returned as an adult and bred at a second nest. In June 2009 a pair produced three young at Kielder Water; these are the first to breed at Kielder for over 200 years." [5]

12.3 Biblical References

It is not possible to identify the Osprey beyond reasonable doubt in the Christian Bibles, as the Hebrew word *raham* occurs only twice namely in Lev 11:18 and Deut 14:17, which are the lists of "unclean" birds. We therefore have no further clues by means of the context as to what exactly is meant. In Deut 14:17 the noun is in the female gender. Some translations e.g. JB, NIV, and A33 translate the Hebrew as "osprey." Koehler & Baumgartner (p. 886) prefer the Egyptian Vulture *(Neophron percnopterus)*. I would go along with the latter as the primary diet of the Osprey would make it less abhorrent as the Egyptian/Carrion Vulture. The translators of A83 changed it from *Visvalk* meaning "osprey" in the 1933 translation to the *Swaan* which means "swan" in their 1983 translation. The LB connects the Osprey to the Hebrew *oznijah*.

5   Wikipedia contributors. "Ospreys in Britain." Wikipedia, The Free Encyclopedia. Wikipedia, The Free Encyclopedia, 30 Jul. 2011. Web. 6 Aug. 2011.

CHAPTER 13
ORDER: GALLIFORMES

> And God said, "Let the waters bring forth abundantly the moving creature that hath life, and fowl that may fly above the earth in the open firmament of heaven."
>
> **Gen 1:20 (KJV)**

13.1 Fowl-like Birds

The relatively large Order *Galliformes* (about two hundred and fifty to three hundred species) are the fowl-like birds. The Latin word *gallus* which means "cock" has led to the name of the order. The order is divided into eight families namely:

- *Phasianidae*   (Chicken, Quail, Partridges, Pheasants).
- *Megapodiidae*   (Mallee Fowl and Brush-turkeys).
- *Cracidae*   (Chachalacas and Currasows).
- *Odontophoridae*   (New World Quails).
- *Numididae*   (Guineafowl).
- *Tetraonidae*   (Grouse).
- *Mesitornithidae*   (Mesites).
- *Meleagrididae*   (Turkeys).

"These birds all are heavy-bodied ground-feeding domestic or game birds. Common names for these birds are "Gamefowl" or "Gamebirds", "Landfowl", "Gallinaceous birds", or "Galliforms." Also "Wildfowl" or just "Fowl" are also often used for *Galliformes*, but usually these terms also refer to Waterfowl (*Anseriformes*), and occasionally to other commonly hunted birds."[1]

---

[1]   Wikipedia contributors. "Galliformes." Wikipedia, The Free Encyclopedia. Wikipedia, The Free Encyclopedia, 9 Jul. 2011. Web. 6 Aug. 2011

All the families in the order have been closely associated with man throughout recorded history. There are at least two reasons:

White-winged Ring-necked Pheasant
(Phasianus colchicus mongolicus)
(Photo: Gary Noon) [2]

- As opposed to sparrows and even doves and pigeons, they are relatively large birds that provide a protein-rich source of food to their predators and to humans. All of them have light, succulent breast meat with a pleasing flavor. This is mainly because they have well developed breast muscles. (According to KFC, the drumsticks are not bad either).
- Because they lay relatively large and highly nutritious eggs, they all have the characteristics of the domestic chicken.

Various species from this order were amongst the earliest, if not the first, birds taken from the wild and domesticated by man. Their well developed breast muscles allow them to become airborne fast and enable them to fly short strong bursts. However, they are habitually ground birds that cannot fly well. Typically they will fly less than one hundred meters. They prefer to run away when endangered. In addition, the cryptic (concealing) paintings on their backs assist them in camouflaging themselves by turning their backs to the danger and running into the cover, for they also have relatively short and strong legs. They are mainly granivorous and herbivorous but are also slightly omnivorous. They feed on an assortment of seeds, leaves, flowers, soft stems, flower- and leaf buds, tubers, rhizomes, bulbs and roots, which they find just above or below ground level. The latter is found by their typical scratching around in search of food. Because of their diet, they are excluded from the "unclean" birds mentioned in modern Bibles. They have large heavy feet, with three toes in front and a shorter one pointing backwards. Many of the species have spurs, making them suitable for cock-fighting.

---

[2]   This image is a reproduction of the photograph taken by Gary Noon and used with permission. Originally placed on www.flickr.com, original photograph can be viewed at: http://www.flickr.com/photos/85245185@N00/454189539/.

13.2 Non-Biblical Families

- *Megapodiidae*: The Megapodes, (nineteen species), (from the Greek, meaning "large foot"), are also called the "mound builders." They are a group of queer Australian and Malayan fowl, and are the only animals above the reptiles in the evolutionary scale that do not depend on their body heat to incubate their eggs. This is done by raking together a huge heap (almost two meters high and six meters wide) of vegetation and soil on the forest floor. The rotting material, like in a compost heap, generates heat up to about thirty-nine degrees Celcius which is used to incubate the eggs that are laid in the mould. The female digs a hole of about one meter deep into the mould, deposits an egg and covers the hole up again. She returns the next day to deposit the next egg until five to eight eggs are laid. The parents are extremely sensitive to the amount of heat generated, and should they feel that too much heat is generated the female would dig down to the eggs to air them for a while, and then cover them up again.
- *Cracidae*: The family consists of two subfamilies namely *Penelopinae*: the Chacha-lacas (twelve species), and Guans (twenty-four species), and *Cracinae*, in other words the Curassows (fourteen species). They are essentially non-migratory forest birds, and the only arboreal family in this order. All are found in the Neotropical Region, and are thus not known to the Old World (Europe, Africa, and Asia) and thus Palestine.
- *Meleagrididae*: The Turkeys are large terrestrial birds found in North and Central America. There are two species namely: The Wild Turkey *(Meleagris gallopavo)*, and the Ocellated Turkey *(Meleagris ocellata)*.
- *Tetraonidae*: Comprising the Grouse (seventeen species).
- *Odontophoridae*: Comprising the NW Quails (thirty-two species).
- *Numididae*: Comprising the Guineafowls (six species).

CHAPTER 14
SUB ORDER: PHASIANOIDEA
FAMILY: PHASIANIDAE

> "And Peter remembered the word of Jesus, which said unto him,
> 'Before the cock crow, thou shalt deny me thrice.'
> And he went out, and wept bitterly."
>
> **Mat 26:75 (KJV)**

14.1 Introduction

We now get to a large family in the Order *Galliformes*, called *Phasianidae*, which are distributed throughout much of the Old World (Europe, Africa, and Asia). They are small, like the Asian Blue Quail *(Coturnix chinensis)* which weighs as little as forty-three grams, to large, like the Indian Peafowl *(Pavo cristatus)* which weighs up to six kilograms. They have the following characteristics:

- They are typically terrestrial birds.
- They have squat bodies that are large in relation to their heads.
- They have short bills, necks, wings and tails.
- Save for a few exceptions, the males are usually larger than the females.

The chicks are precocial, in other words, their eyes are open when they hatch and from day one they require relatively little parental care and can instinctively find food by themselves. However, when still very young, they are led to food by the parent(s). In many species the chicks resemble the female at first. They live in a wide range of habitats, from the lowlands to the high alpine meadows in both tropical and temperate climates.

The Chukar Partridge or Chukar *(Alectoris chukar)* for example can be found within the parks of Jerusalem, e.g. the Wolfson Garden, and the Sherover Promenade area. One can find them feeding in open areas in the early morning, and retreating to patches of scrub, long grass or rocky ledges during the heat of the day. One can find them again in the late afternoon, foraging near some open water. In the country they are seen working their way up steep slopes of foothills when foraging. Those living on mountain slopes, like those found on the Golan Heights, occupy the higher zones during the summer. They descend in the autumn to escape the adverse winter weather conditions of the high altitudes. Another common sight in the Middle East from this family, is the Black Francolin *(Francolinus francolinus francolinus),* as it has adapted well to the agricultural environments created by the Kibbutzim. When driving through the fields on a kibbutz, they are a common sight along the road. Clutches vary from seven to twelve eggs. Incubation is by the female only, taking eighteen to twenty-five days. After hatching, the chicks are covered in down, with the primaries appearing during the following three to four weeks. Most species become sexually mature within a year.

This large family is subdivided into two subfamilies: *Perdicinae:* the Partridges and their allies (one hundred and six species: including forty Francolin and fourteen Quail species), and *Phasianinae:* the Pheasants (forty-nine species).

14.2 Partridges

> "Like a partridge that hatches eggs it did not lay,
> is the man who gains riches by unjust means."
>
> **Jer 17:11 (NIV)**[1]

Although partridges are found in the Mediterranean, e.g. the Barbary Partridge *(Alectoris barbara),* the Red-legged Partridge *(Alectoris rufa),* the Grey Partridge *(Perdix perdix),* and the Rock Partridge *(Alectoris graeca),* it is really only the Chukar that is common in the Middle East. The Arabian Partridge *(Alectoris melanocephala),* and Philby's Partridge *(Alectoris philyi),* are restricted to small areas in Arabia. On the steep, rocky slopes of the desert and semi-desert areas, one finds the Sand Partridge *(Ammoperdix heyi).*

14.2.1 Biblical References

The Hebrew word *qore',* used in 1 Sam 26:20 is associated with the partridge by Koehler & Baumgartner (p. 851), which fits in well with the typical habitat of the species. Douglas, (p.

---

156), mentions that the Rock Partridge (*Alectoris graeca*) is regularly hunted in many parts of the Middle East and southeast Europe, but regarding the Middle East, this does not correspond with the present distribution of the specific species.

It might not be possible to determine the exact species meant, but in Jer 17:11 we find a proverb accurately describing a behavioral pattern of the *Phasianidae* family that, although mainly monogamous, instances of successive bigamy have been reported. Koehler & Baumgartner substantiate their

Sand Partridge *(Ammoperdix heyi)*
(Photo: Dr. Marcel Holyoak) [2]

claim by quoting an observation by Aharoni of two layings of eleven eggs each of two different females in the same nest.

Another alternative in this regard would be the family of Sandgrouse *(Pteroclidae)* of which five species are resident in the Middle East, namely: the Pin-tailed Sandgrouse *(Pterocles alchata)*, Spotted Sandgrouse *(Pterocles senegalus)*, the Crowned Sandgrouse *(Pterocles coronatus)*, Lichtenstein's Sandgrouse *(Pterocles lichtensteinii)*, and the Black-bellied Sandgrouse *(Pterocles orientalis)*.

14.3 Quail

> "That evening quail flew in and covered the camp."
> **Ex 16:13 (MSG)** [3]

The Common Quail *(Coturnix coturnix),* as a rule avoids being on bare soils. They prefer dense bushes, shrubs, and herbage less than one meter tall, including areas where crops

---

[2]   This image is a reproduction of the photograph taken by Dr. Marcel Holyoak and used with permission. Originally placed on www.flickr.com, the original photograph can be viewed at: http://www.flickr.com/photos/maholyoak/5917426425/.

[3]   Scripture taken from The Message. Copyright © 1993, 1994, 1995, 1996, 2000, 2001, 2002. Used by permission of NavPress Publishing Group.

Common Quail *(Coturnix coturnix)*
(Photo: Bo Jonsson) [4]

are cultivated. Most of the galliform birds are more or less resident, but some of the smaller temperate (between the tropics and the polar circles) species such as the Common Quail do migrate over considerable distances. Our interest lies in the migration from the Western Palearctic across the Mediterranean to the sub-Sahara region.

The Common Quail is generally reluctant to fly, but once they are airborne, they fly swiftly and directly to their destination. This enables them to cover large distances, including flying over lakes, the sea, and making desert crossings. They typically fly only a few meters above the water and/or dunes. However, the weight of the bird makes it vulnerable to the strength and direction of the prevailing wind. It is known that after experiencing adverse wind conditions in flight, they are so exhausted after crossing a large body of water that they are forced to land and rest as soon as they reach the opposite shore. It is then that they are unable to flee away from danger, making harvesting them and/or preying on them, an easy task.

In days gone by, they migrated in phenomenal, almost unbelievable numbers. One researcher has, from the Biblical accounts, estimated that some nine million birds were killed by the Israelites in the Sinai Peninsula. Even as late as the 1920's exports from Egypt numbered millions per annum. However, by the 1930's these populations were almost completely decimated by the uncontrolled over-exploitation.

14.3.1 Biblical References

Today it is widely accepted that the Hebrew word *slaw* refers to the Common Quail *(Coturnix coturnix)*. In the Bibles they are best known as a food source for the Israelites in the Sinai Peninsula as described in Ex 16 and Num 11 and referred to in Ex 16:13 and in Num 11:31. In Ps 105:40 and Ps 78:27 the Psalmists are referring to this natural wonder, with the latter referring only to "birds" in general. It is interesting to note that wonders in the Bibles often are quite natural occurrences. In this case the time, location,

---

[4]   This image is a reproduction of the photograph taken by Bo Jonsson and used with permission. Originally placed on www.flickr.com, the original photograph can be viewed at: http://www.flickr.com/photos/35040514@N07/3268628935/sizes/l/in/photostream/.

and extent of the natural wonder have been engineered by the God of Israel to prove his provision to his people.

14.4 Red Junglefowl

> "O, Jerusalem, Jerusalem. How often have I longed to gather your children together, as a hen gathers her chicks under her wings, but you were not willing."
>
> **Luke13:34 (NIV)[5]**

"The crow of the domestic or village cock is one of the most familiar natural noises the world over. What may not be such common knowledge is the fact that the Red Junglefowl *(Gallus Gallus),* most likely the wild ancestor of the domestic chicken, has an extensive vocal repertoire." (del Hoyo, et al, Vol. 2, p. 445).[6]

A thirteenth century Arabic writer states that "it knows the watches of the night", which Peter, one of the disciples of Jesus Christ, could confirm. It was present in Egypt by 1500 BC, which coincides with the Israelite slavery of about 1700–1400 BC. It is believed to be introduced to Palestine proper by about 200 BC. To this day, one is able to bring an early morning visit to the Church of the Renunciation, just outside the Lion's Gate in Jerusalem, and as the day breaks, hear a cock crowing in a nearby Arab suburb. It is referred to in the *Talmud* and has been a traditional Jewish atonement offering at the feast of *Yom Kippur* (Day of Atonement).

Because the junglefowl has been part of human life seemingly for ever, it is quite impossible to determine when the first junglefowls were domesticated by humans. Some anthropologists estimate that it happened somewhere in the Bronze Age round about 4000 BC. Other evidence shows that they were domesticated in the Indus Valley in India in about 3200 BC. As a lightweight and therefore portable source of food, they accompanied voyagers everywhere, arriving in Europe for example by 1500 BC. Magellan, Cook, and Columbus took them along on their journeys but often found them already domesticated by the locals in the far off places they arrived at.

---

[5]   THE HOLY BIBLE, NEW INTERNATIONAL VERSION®, NIV® Copyright © 1973, 1978, 1984, 2011 by Biblica, Inc.™ Used by permission. All rights reserved worldwide.

[6]   del Hoyo, J., Elliot, A. & Sargatal, J. eds. (1994). Handbook of the Birds of the World. Vol. 2. New World Vultures to Guineafowl. Lynx Edicions, Barcelona. Used with permission.

Red Junglefowl *(Gallus Gallus)*
(Photo: Nikita Hengbok) [7]

During its long history in close proximity to humans, the Junglefowl has had an impact on various aspects of human life. It is virtually impossible to overestimate its impact on our daily lives, and its influence is probably unsurpassed by any other animal or bird. The horse would probably be its closest rival in this respect.

There are few people who have not had some contact, albeit indirect, with one of the junglefowl species. Whilst continuously being exploited throughout known history, it has at the same time been admired, feared, viewed with suspicion, and in addition, used for religious, medicinal, and entertainment purposes wherever humans settled. It even had an influence on languages all over the world. For example in English we find words and phrases like "cock, cockatoo, cocky and cockiness, cockroach, cocktail, cock-a-hoop, cock-brained, cock-eyed, cock-sure, cockpit, chicken-breasted, chicken feed, chicken-heart, chicken-pox, chicken out, etc."

14.4.1 Biblical References

In Job 38:36 the Hebrew *sekwi* refers to a cock. This is the only occurrence of the word, making it problematic. The first half of this verse contains an unexplained word, which is translated as an "ibis" in A83, JB and TEV.

Some translations like the RSV, A83, and TEV, include the word "cock" in Prv 30:31. Within the context of the verse it makes sense, but the Hebrew is not absolutely clear. A situation that is duely acknowledged in TEV.

In the New Testament the cock, (Greek: *alektor*), figures prominently in the denial of Jesus by Peter, as described in Mat 26:34, 75, 75; Mrk 14:30, 68, 72; and Luk 22:34, 60, 61. In Mrk 13:35 the crowing of the cock is used by Jesus as a symbol of uncertain

---

[7] This image is a reproduction of the photograph taken by Nikita Hengbok and used with permission. Originally placed on www.flickr.com, the original photograph can be viewed at: http://www.flickr.com/photos/hengbok-nikita/6027368025/in/set-72157626138757633.

times. From one of the Lord Jesus' most poignant similes, in Mat 23:37 and Luk 13:34, it is clear that Jesus and the people of his time had closely observed the behavior of a hen (Greek: *ornis*) and her chicks, be it in the wild or those that have been domesticated.

14.5 Indian Peafowl

> "The king had a fleet of ocean-going ships at sea with Hiram's ships. Every three years the fleet would bring in a cargo of gold, silver, and ivory, and apes and peacocks."
>
> **1 Kng 10:22 (MSG)**[8]

The Indian Peafowl or Blue Peafowl (*Pavo cristatus*), is a large and brightly colored bird originally a native of South Asia. Later it was introduced and became semi-feral in many other parts of the world. The male, known as a "Peacock", is predominantly blue with a fan-like crest of spatula-tipped wire-like feathers. He is, however, best known for his long train (tail) made up of elongated upper-tail covert feathers which bear colorful eyespots. These stiff and elongated feathers are raised into a fan and quivered in a spectacular display during courtship. The train of the male takes four years to develop fully, reaching a length of up to one and a half meters. The female, known as a "Peahen", lacks the elaborate train, and

Indian/Blue Peafowl *(Pavo cristatus)*
(Photo: Gary Knrd) [9]

---

[8] Scripture taken from The Message. Copyright © 1993, 1994, 1995, 1996, 2000, 2001, 2002. Used by permission of NavPress Publishing Group.

[9] This image is a reproduction of the photograph taken by Gary Knrd and used with permission. Originally placed on www.flickr.com, the original photograph can be viewed at: http://www.flickr.com/photos/avianphotos/5273355300/.

she has a greenish lower neck and has an overall duller brown plumage. They are found mainly on the ground in open forest or cultivated areas where they forage for berries, grains, and bits of other vegetation. They will also prey on small snakes, lizards, small rodents, and the like. Their loud calls make them easy to detect, and in forest areas, often indicate the presence of a large predator such as a tiger. They forage on the ground, moving in small groups and will usually try to escape from danger on foot through undergrowth and so avoid flying. They will fly up into tall trees to roost, however. It is not surprising that the Indian Peafowl is the national bird of India.[10]

"By and large not used by man as a staple source of food, not widely used for medicinal purposes or sport or hardly ever viewed with fear or superstition, the Indian Peafowl *(Pavo cristatus)*, has mostly been admired for the beauty of the male." (del Hoyo, et al, Vol. 2, p. 467).[11] In the latter part of the fourth century BC, Alexander the Great brought the bird to Greece, and was so impressed by the beauty of the male, that he imposed a severe penalty on anybody harming it. Aristotle quotes Aldrovandros, who wrote: "though he is a most beautiful bird to behold, yet the pleasure of the eyes is compensated with many ungrateful strokes upon the ears, which are often afflicted with the odious noise of his horrid cry." (del Hoyo, et al, Vol. 2, p. 467).[12] The Peafowl is considered to be sacred by the Hindus in its Indian homeland, and outside of India it has come to symbolize wealth and power.

14.5.1 Biblical References

Twice in the Old Testament, the Hebrew word *tukkijjim* is used, namely in 1 Kings 10:22 and 2 Chr 9:21, listing some of the importations made by Solomon. In both cases the word *tukkijjim* is preceded by the word *qophim* translated as "apes." Older translations e.g. LB and A33, translated these two words as "apes and peacocks", as the beauty of the bird would fit in with the splendor of Solomon's kingdom. For this argument the Green Peafowl *(Pavo mutucus)*, or even one or more of the Pheasants (e.g. *Chrysolophus Spp.)* would be fitting alternatives. In addition, we should keep in mind that Solomon ruled from about 971–931 BC, which would place this Old Testament occurrence six hundred

---

10 Wikipedia contributors. "Indian Peafowl." Wikipedia, The Free Encyclopedia. Wikipedia, The Free Encyclopedia, 25 Jul. 2011. Web. 6 Aug. 2011.

11 del Hoyo, J., Elliot, A. & Sargatal, J. eds. (1994). Handbook of the Birds of the World. Vol. 2. New World Vultures to Guineafowl. Lynx Edicions, Barcelona. Used with permission.

12 del Hoyo, J., Elliot, A. & Sargatal, J. eds. (1994). Handbook of the Birds of the World. Vol. 2. New World Vultures to Guineafowl. Lynx Edicions, Barcelona. Used with permission.

years before the time of Alexander the Great (356–323 BC), which would pose the question why somebody like Aristotle did not know or, if he knew, mention the existence of such a bird in the Middle East. At first it was though that *Tarshish* meant a Phoenician center somewhere in Spain, but reasonable consensus has been reached that it refers to a type of ore-bearing ship, and not to their destination. Trading was Solomon's forte, and contact with India and its peafowl cannot be excluded. But in the absence of undisputed evidence, we have to agree with Douglas (p. 156) when he states that "there is no independent evidence to confirm the identification of *tukkijjim*, though it is suggested that this word is derived from the Singhalese *tokei*, meaning "peacock." Later translations e.g. A83, NIV, and JB have therefore translated the words as "apes and baboons", opting for a grouping of two primate families. In concluding, we have to concede that we simply do not know beyond a reasonable doubt which bird or even animal the author had in mind.

CHAPTER 15
ORDER: STRIGIFORMES

> "I'm like a buzzard in the desert, a crow perched on the rubble.
> Insomniac, I twitter away, mournful as a sparrow in the gutter."
>
> **Ps 102:7–8 (MSG)**[1]

15.1 Nocturnal Raptors

The order *Strigiformes* consists of the nocturnal (and solitary) birds of prey, collectively known as the Owls. Amongst others they possess the following characteristics:

- Large forward-facing eyes and ear-holes;
- They have a flat face; and usually a conspicuous circle of feathers, a *facial disc*, around each eye.
- Owls can rotate their heads and necks as much as 270 degrees in either direction.
- They are farsighted and therefore unable to see clearly anything within a few centimeters of their eyes. Caught prey can be felt with the use of filoplumes — small hair-like feathers on the beak and feet that act as "feelers."
- They have strong, hawk-like bills and a cere (membranous covering) at the base of the bill.
- They have strong, sharp, and hooked talons. This prompted Linnaeus in 1758 to place the owls and hawks in the same group. In 1890 it was noted that owls share many characteristics with the Nightjars *(Caprimulgidae),* a view that gained universal acceptance to this day.

Owls of all kinds have fascinated humans since ancient times. They are either feared (Africa) or venerated (Australia), despised (e.g. as "unclean" in the Bibles), or admired as wise. Biblical evidence confirms this with no less than twenty-five references to nine

---

[1]   Scripture taken from The Message. Copyright© 1993, 1994, 1995, 1996, 2000, 2001, 2002. Used by permission of NavPress Publishing Group.

species in the Old Testament (refer to Chapter 36), and usually associated with desolation, destruction and ruin. Paintings of owls on cave walls in France date back fifteen to twenty thousand years.

Owls generally have loud vocal voices because they are largely nocturnal birds and therefore cannot rely on visual communication. Over the ages this, together with their "human-like" faces (with large forward-facing eyes) and "ears" have contributed to the fact that across the globe they are associated with a wide range of totally absurd superstitions. This is manifested in a wide range of realms such as magic, sorcery, witchcraft, prophesying and soothsaying, birth, death and dying, and even the weather. The family name *Strigidae* is from *strix*, from the Greek meaning "witch", transformed to *striga* in Latin.

If a Eurasian Scops-owl *(Otus scops)* called near the house of a sick man in Sicily, legend has it that he would die three days later. To the Chinese, owls are "dig-

Burrowing Owl *(Athene cunicularia)*
(Photo: Dori (dori@merr.info) CC-BY-SA 3.0) [2]

ging the grave", snatching away souls. In the Arab world owls are regarded as the souls of the dead. In Hispaniola (Haiti and the Dominican Republic), Owls are regarded as having supernatural powers and could transform themselves into witches. Indian tribes living on the pampas of Argentina fear the Burrowing Owl *(Athene cunicularia)*.

In Africa owls have a profound influence on the superstitious believes of the local black people, and consequently, in all African cultures little mercy is shown towards them. In contrast, amongst the Aboriginal people of Australia, all "birds of evil omen" were protected by the woman folk, as the kinship between each woman and the lives

---

[2] This image is a reproduction of the photograph taken by Dori (dori@merr.info) CC-BY-SA 3.0 and used with permission. Originally placed on www.flickr.com, the original photograph can be viewed at: http://www.flickr.com/photos/liroi/4654445400/in/set-72157608183102340.

of the other females in her family were connected to these birds, and it was not known which owl guarded which person.

Owls have since ancient times been associated with wisdom. For confirmation of this, one only has to visit one's local kindergarten. The association between the Greek goddess of wisdom, Athene, and the Little Owl *(Athene noctua),* has been perpetuated in it's the generic name as well as three other species in this genus. As far back as 450 BC a Little Owl was depicted on a coin in Greece. They have been associated with libraries and universities, with a Little Owl coin being minted in conjunction with the 350-year anniversary of the University of Helsinki.

Today, many people are totally obsessed with owls. Known as *Owlaholics* they would collect whatever depicts an owl: figurines, posters, paintings, matchboxes, cards, stamps, and what not, or have as many everyday utensils in owl motifs as possible. People preferring the night-time are called "night owls." In Italy a police patrol car is called a *civetta* meaning a "small owl", and in Bologna a policeman on the beat is known as a *gufo* or owl.

Parts of owls or their eggs featured in various folk remedies. For example:

- For better eyesight: eat the "awake" eye of an owl, which is said to float on water as opposed to the "sleepy" eye that sinks; or alternatively, charred and powdered owls' eggs.
- Owls' egg soup, is reckoned effective against epilepsy, however, it should be prepared when the moon is on the wane.
- To eradicate a love for liquor: Owl eggs cooked in wine for three days. Alernatively: salted owl or an omelette of five, nine or thirteen owls' eggs (France).
- For rheumatism: burning owl feathers over charcoal, or eating baked owl (Poland).
- The Puerto Rican Screech-owl *(Otus nudipes)* can cure asthma as it eats coffee beans.
- Owl soup in moderate doses served as a remedy for whooping cough (Yorkshire).
- The body parts of owls e.g. Collared Scops-owls *(Otus lettia)* feature in traditional Chinese medicines. (China, Korea and Thailand).
- To compensate for a scarcity of owls to control rodents, or to avert bad weather, a dead owl is hanged on or above the barn door in parts of England.

Through the ages, from ancient Babylon to modern day China, owls have appeared on the menu in various parts of the world. For example:

- Eskimos describe the Snowy Owls *(Nectea scandiaca)* as "lumps of fat."
- Burrowing Owls are served as a delicacy to people as a convalescent diet in Uruguay in order to recover health and strength gradually after sickness or weakness.
- But South-African blacks maintain that owl flesh contains scurf (dandruff), causing it to smell of death and thus inedible.

Since about the eighteenth century, owls have been widely appreciated as eradicators of rodents and other pest species on farms. Attempts to utilize the Common Barn-owl *(Tyto alba)* as a biological control measure against rodents on small islands like the Seychelles, Hawaii, and Lord Howe Island (between Australia and New Zealand) have however been unsuccessful and often proved to be detrimental to the local species.

15.2 Classification

There are two families, five subfamilies and six tribes in this Order.

The family *Tytonidae* consists of two subfamilies:

- *Tytoninae*, also known as the Tyto Owls, consists of fourteen species:
    - Sooty-owls, (two species). They are rainforest specialists, who have the largest eyes for hunting in complete darkness.
    - Masked-owls, (six species), including several isolated, poorly known island endemics.
    - Typical Barn-owls, (four species). Refer to Chapter 16 for a detailed discussion.
    - Grass-owls, (two species), occurring widely in the Old World (Europe, Africa, and Asia) tropics, reaching South-Africa in the west and southern Australia in the east. They are rodent specialists (seventy-six to ninety-eight percent of their diet in Africa).
- *Phodilinae,* (two species), namely: the Oriental Bay-owls and Congo Bay-owls. Their facial discs are divided at the forehead, and both species are forest specialists.

The family *Strigidae* (Typical Owls) consists of three subfamilies:

- *Striginae*, consisting of three tribes, namely:
    - *Otini*: Scops-owls (forty-five species), and the Screech-owls (twenty-three species).
    - *Bubonini*: Eagle-owls, and Allies (twenty-five species).
    - *Strigini:* Wood-owls (twenty-four species).
- *Asoninae*, also known as the Eared-owls (nine species).
- *Sumiinae*: consists of three tribes namely:
    - *Surniini*: Pygmy-owls and Owlets (thirty eight species).
    - *Aegoliini*: Saw-whet Owls (four species).
    - *Ninoxini:* Hawk-owls (twenty one species).

CHAPTER 16
FAMILY: TYTONIDAE

> "the little owl, the great owl, the white owl,"
> **Deut 14:16 (NIV)**[1]

16.1 Common Barn-owl

The Common Barn-owl *(Tyto alba)* is called in this way to clearly distinguish it from other species in the Barn Owl family *Tytonidae*. There are at least twenty-eight sub-species.

"It is easy to understand the long-established fascination that human beings have for owls. This ranges across all cultures, and is perhaps most widely appreciated with regard to the Common Barn-owl. This species' ghostly appearance and silent flight, the loud and eerie shrieking and hissing sounds it makes, and its largely nocturnal habits have earned it a myriad of names in the folklore of many languages. These often betray an association in the human mind with witchcraft, magic and death, but in some areas the species is considered a lucky omen." (del Hoyo, et al, Vol. 5, p. 63).[2]

It is therefore not surprising that the Common Barn-owl has no less than twenty-two other common names in English, e.g.: Barnyard Owl, Cave Owl, Church Owl, Delicate Owl, Demon Owl, Death Owl, Dobby Owl, Ghost Owl, Golden Owl, Hissing Owl, Hobgoblin or Hobby Owl, Monkey-faced Owl, Night Owl, Rat Owl, Stone Owl, Scritch Owl, Screech Owl, Silver Owl, Straw Owl, White Owl, and White Breasted Owl. These names may refer to the appearance, the call, the habitat or the eerie, silent flight of the bird. "Golden Owl" might also refer to the related Golden Masked Owl *(Tyto aurantia)*. "Hissing Owl" and, particularly in the USA, "Screech Owl", refer to the piercing calls of these birds. The latter name, however, more correctly applies to a different group of birds, the screech-owls in the genus *Megascops*. "The Barn Owl's scientific name, estab-

---

[1] THE HOLY BIBLE, NEW INTERNATIONAL VERSION®, NIV® Copyright © 1973, 1978, 1984, 2011 by Biblica, Inc.™ Used by permission. All rights reserved worldwide.

[2] del Hoyo, J., Elliot, A. & Sargatal, J. eds. (1999). Handbook of the Birds of the World. Vol. 5. Barn-Owls to Hummingbirds. Lynx Edicions, Barcelona. Used with permission.

lished by G.A. Scopoli in 1769, literally means 'white owl', from the onomato-poetic Ancient Greek *tyto* for an owl, (compare English 'hooter'), and Latin: *alba*, 'white'."[3] Onomatopoeia is a word that imitates or suggests the source of the sound that it describes.

"The Common Barn-owl has the distinction of being one of the world's most widely distributed land birds, as well as the most intensively studied of all owls, particularly in Europe and North America. However, of the twenty-eight subspecies presently recognized, most are poorly known, and some isolated populations may represent separate species." (del Hoyo, et al, Vol. 5, p. 36).[4]

Morphologically the Tyto Owls are characterized by the following features:

Common Barn-owl *(Tyto alba)*
(Photo: Peter G. Trimming) [5]

- Exceptionally large rounded heads.
- Heart-shaped facial discs, which act as sound-amplifying reflectors.
- Narrowly tapered bodies for improved aero-dynamics.
- Short squared tails.
- Broad and long wings (resulting in a high surface area to body-mass ratio. This increases lift and allows slow flight and sharp turning without stalling).
- Long legs, strong feet, and well developed talons, all resulting in a wide spread at the point of contact.

---

3   Wikipedia contributors. "Barn Owl." Wikipedia, The Free Encyclopedia. Wikipedia, The Free Encyclopedia, 6 Aug. 2011. Web. 6 Aug. 2011.

4   del Hoyo, J., Elliot, A. & Sargatal, J. eds. (1999). Handbook of the Birds of the World. Vol. 5. Barn-Owls to Hummingbirds. Lynx Edicions, Barcelona. Used with permission.

5   This image is a file from the Wikimedia Commons. Full details of the file can be found at: http://en.wikipedia.org/wiki/File:Tyto_alba_-British_Wildlife_Centre,_Surrey,_England-8a_(1).jpg. This image is licensed under the Creative Commons Attribution 2.0 Generic license by Wikipedia Commons user Peter Trimming from Croydon, England. Flickr user name is "Peter G Trimming".

Common Barn-owls are, in many ways, extremely well adapted birds, specifically for nocturnal (night-time) hunting. The heart shape of the facial disc forms two concave sound wave troughs to the ears. Their external ears are placed asymmetrical, with the left ear slightly higher than the right. This allows them to determine the location of sounds much more accurately. Their flight feathers are adapted in three ways resulting in absolute silent flying. This reduces the chances of prey detecting them, and in addition assists them in locating prey sounds when hunting in flight.

> "The pride of the peacock is the glory of God."
> "The crow wishes everything was black, the owl, that everything was white."
>
> **William Blake (1757-1827)**
> *(English Visionary mystic, Poet, Painter, and Engraver)*

When in a state of rest, the Common Barn-owls typically hang their plumage loose around their bodies, like a man wearing an over-sized coat. They would stand on one leg with the body leaning forwards. The facial disc is flattened, and the eyes closed. The behavior when they are threatened in any way is even more typical. They bow down towards the intruder, they spread their wings and tail, and the head is lowered and swayed slowly from side to side. To create an even more formidable image, the above is accompanied by much hissing and bill snapping and even sudden forward lunges towards the threat. The sounds produced by the tytonid owls have been grouped into five to twelve types of sounds, with the eerie shrieking and hissing sounds universally known.

All owl species regularly regurgitate (vomit) a soft ball of undigested food, bones, and hair. This is commonly known as a "pellet' or "owl pellet." Because the stomach juices found in the Common Barn-owl are less acidic than those of most other birds of prey, even small and delicate bones are preserved intact in these pellets. Therefore, we are fortunate in that by analysing the pellets it enables us to identify the range of species that the bird had been preying on. The diet of the Common Barn-owl consists primarily (seventy-five to one hundred percent), of small to medium sized mammals, with most of them being rodents like mice, rats, and voles. Where these are scarce or completely absent, the Common Barn-owls are able to adapt successfully, e.g.: in the Namib Desert of southern Africa, lizards may comprise about half of its diet, followed by scorpions. At the Karnak Temple, in Egypt, about half of its diet consists of Doves *(Columbidae)* and Sparrows *(Passeridae).*

As they are in the order *Strigiformes,* Common Barn Owls usually hunt alone. They drop down onto the prey from a perch, or find prey by a low (about three meters above ground level) searching flight, including the occasional hovering. They are able to hunt efficiently in what is, to the human eye, total darkness. However, if necessary, they would hunt in daylight as well. When hunting during the day they are often mobbed by smaller birds. Common Barn Owls consume more rodents than possibly any other creature on earth. A nesting pair and their young can eat more than a thousand rodents per year. This makes the Common Barn Owl one of the most economically valuable wildlife animals to farmers, and all they ask for in return is a safe and suitable nesting site. Farmers often find these owls more effective than poison in keeping down rodent pests. Barn Owls are encouraged to move in and stay by providing them with suitable nest sites in the form of "nest boxes" or other natural and/or man-made structures.

Common Barn-owl *(Tyto alba)*
(Photo: Rus Koorts) [6]

Their breeding activities start about six to eight weeks before they actually start laying the eggs. This includes food presentation by the male to the female which is an important part of courtship behavior. The Common Barn-owl is a cavity-nester. Typically they build their nests in trees, in caverns, and suitable man-made structures like barns, attics, clock towers, unused mine shafts, etc. They are site-faithful, returning year after year, provided that little or no disturbances occur that would scare them away. An extreme example is known. Calculations according to the layers of prey remains in a cave site, show that Greater Sooty-owls *(Tyto tenebricosa)*, had been using a site for ten thousand years.

Clutches vary from four to seven eggs, and are incubated by the female for about thirty days. Hatching of the chicks is asynchronous, (the state of not being synchronized), resulting in a size hierarchy amongst the brood, as they hatch one after another. Photographs of a number of chicks in a row, positioned from the smallest to the largest,

---

[6]   This image is a reproduction of the photograph taken by Rus Koorts from Pretoria, South Africa and used with permission. Originally placed on www.flickr.com, the original photograph can be viewed at: http://www.flickr.com/photos/ruslou/5810551781/sizes/l/in/photostream/.

are most popular in galleries. Most of the chicks survive under normal circumstances, with much variation according to the region, breeding season, and annual variations in circumstances. The chicks are fully fledged in about forty days, and reach maturity in about one year.

16.2 Non-Biblical References

Amongst paleontologists (scientists studying prehistoric life), Common Barn-owls are known to have been present in Britain for at least ten thousand years. Their skeletons were found among the remains of an Iron Age (from about 1200 BC) village at Glastonbury in the south-west of England. Because of their distinctive features, Common Barn-owls are easily identified on wall paintings and bas-reliefs from ancient Egypt. In addition, it is found to be among the six or more species that have been positively identified from mummified remains found in tombs. Owls are to this day found frequenting tombs and temple sites, which probably strengthened the believe by some Egyptians that a person had three souls. One of them, which were named *"Ba"*, was in the form of a human-headed bird that never left the burial site. The owl symbol is used in hieroglyphics as the "m" sound or to indicate the bird itself.

In pre-Columbian America the owl motif is widespread. These motifs even indicate the recognition by the local people of the owl's beneficial association with man as an effective means of rodent control. Owls are depicted in the hieroglyphics of the Mayan Indians of Central America, and the Zapotec of southern Mexico, associated mainly with death. The latter believing that the Common Barn-owl is commissioned to give notice of eminent death and is also responsible for fetching the soul of the deceased. In addition, the owl would symbolize courage, strength, or wisdom in other indigenous New World (the western hemisphere, including the Americas) cultures.

16.3 Biblical References

The Hebrew word *tinshemet*, from the stem *nasham*, "to pant, breath", is found on three occasions in the Old Testament. In the lists of "unclean" birds, the word is found in Lev 11:18 and Deut 14:16 and it is generally accepted as referring to the Common Barn-owl. The only other occurrence of this word is in Lev 11:30, where a list of "unclean" animals "that move about on the ground", is given, and there it is translated as Chameleon *(Cameleo cameleo)*. The hissing of the Common Barn-owl and the Chameleon when threatened may have led to this particular Hebrew stem being used. Unfortunately this stem is used

in only one other instance namely in Is 42:14 in connection with the panting of a woman in childbirth. In Modern Hebrew the Common Barn-owl has retained the Biblical Hebrew name.

Yellowbilled Stork *(Mycteria ibis)*
(Photo: Rus Koorts) [7]

---

[7] This image is a reproduction of the photograph taken by Rus Koorts from Pretoria, in South Africa and used with permission. Originally placed on www.flickr.com, the original photograph can be viewed at: http://www.flickr.com/photos/ruslou/1421538631/in/set-72157602087187880

CHAPTER 17
FAMILY: STRIGIDAE

> "From generation to generation it will lie desolate;
> No one will ever pass through it again.
> The desert owl and the screech owl will possess it;
> The great owl and the raven will nest there."
>
> **Isaiah 34:10–11 (NIV)**[1]

17.1 Typical Owls

There are one hundred and eighty-nine living species of "Typical Owls", also called "True Owls" grouped together in this family called *Strigidae*. In laymen's terms they are therefore often called the "Strigids." They range in size from the Elf Owl *(Micrathene whitneyi)*, weighing about forty grams, to the Eurasian Eagle-owl *(Bubo bubo)*, and Blakiston's Fish-owl *(Bubo blakistoni)* weighing about four kilograms or more. They are characterized by:

- Relatively large heads and round facial discs that encircle the eyes.
- Cryptic (concealing) plumage. Sexual dimorphism is the condition in which the males and females in a species are morphologically (form and structure) different, as is the case with many birds. Owls tend not to exhibit sexual dimorphism in their plumage. Most owls however exhibit RSD (reversed sexual dimorphism) in size, whereby females are generally larger than males.
- Large external ear openings.
- Large wings (wide and long) in relation to body mass.
- Short, stout bills with sharply hooked upper mandibles with sharp cutting edges.
- A reversible fourth toe. In the case of the Fishing-owls *(Ketupa Spp.* from Asia and *Scotopelia Spp.* from Africa) they have unfeathered tarsi, with the soles of the feet having spicules aiding in holding slippery piscine (fish-like) prey.
- They all have relatively short tails.

- Relatively large (often brightly colored) eyes, some larger than the human eye. The eyes are positioned well apart on the front of the flattened face, but are immobile in the skull. They have large pupils. All of this results in their visual sensitivity in low light being up to a hundredfold better than diurnal birds such as Pigeons and Doves (*Columbiformes*).

Elf Owl *(Micrathene whitneyi)*
(Photo: Greg Lasley) [2]

Because of the immobility of their eyes in the eye sockets, the owls have to turn their heads to see to the sides. To do this they can rotate their heads more than two hundred and seventy degrees. The highest number of owl species (about eighty percent) occurs in tropical forests, but owls also occupy a wide variety of forest and wooded habitats and grasslands. Their habitat choices are determined mainly by the availability of food and suitable nesting sites. The latter is crucial for the *Strigids* as they do not build their own nests, but utilize a wide variety of natural or artificial cavities.

Owls are almost exclusively carnivorous. In other words, they feed on a wide variety of vertebrates as well as invertebrates, ranging from hares and rabbits, to insects and even bats. Even the Fish Owls do not feed exclusively on fish. Generally speaking the size of the prey correlates with the size of the bird, but smaller prey would be taken as tidbits by bigger birds. Food is usually detected by audial clues, and they make use of the "perch and pounce" technique of hunting. Active hunting on the wing does however occur from time to time.

The prey is killed by a rapid forceful squeezing of the talons into the vital organs of the prey. If needed, this is often followed by a bite to the back of the scull of the prey. Small meals are swallowed whole, while bigger items are torn apart and devoured piecemeal. Pellets are regurgitated while perching in the daylight hours, containing fur and undigested bones from the previous meal.

Although eight species are polygamous in times of abundant food supply, *Strigids* are mostly monogamous. The frequency at which they breed and the size of the clutches de-

---

[2]   This image is a reproduction of the photograph taken by Greg Lasley http://www.greglasley.net and used with permission. Originally placed on www.flickr.com, the original photograph can be viewed at: http://www.flickr.com/photos/gwldragon/4604281967/.

pends solely on the extent to which food is available at the time. Vocalizing by the breeding male begins about a month before nesting begins. They will nest in almost any place provided there is little interference and sufficient food supplies. In Southern Africa, the Spotted Eagle-owl *(Bubo africanus africanus)* often nest on the ground, but are generally known to use nests built by the Hamerkop *(Scopus umbretta)* and Social Weavers *(Philetairus socius)*. The smaller *Strigids* use the cavity nests of Woodpeckers *(Picidae)* and Barbets *(Capitonidae)*. Eggs are laid at intervals of forty-eight hours, with a strict division in tasks by the breeding pair. Clutches vary according to the size of the species, from about three eggs for smaller species and about five eggs for larger species. Only the female, who is fed by the male, does the incubation, which lasts for twenty-two to thirty-two days. The chicks are fully fledged in a few weeks.

17.2 Eurasian Eagle-owl

The Eurasian Eagle-owl *(Bubo bubo)* is often called Europe's largest owl. It size ranges from about fifty-eight to about seventy-five centimeters long, and weighs about two to four kilograms. The Eurasian Eagle-owl is also called the Common-, Great-, or Northern Eagle-owl. More than fourteen subspecies have been identified and described. In the Middle East the Eurasian Eagle-owl appears to intergrade with the Pharaoh Eagle-owl *(Bubo ascalaphus)*. It is described as an almost barrel-shaped bird. Its prominent ear-tuffs and orange eyes make it one of the more distinctive species. Typical of the family that it belongs to, it has powerful feet and a strong bill. It prefers locations with little human presence and/or activity, and is often found topographically inaccessible terrain, such as undisturbed wilderness areas with virgin forest and deep gorges, and also semi-desert and desert areas with many rocky cliffs, caves and ravines. Its diet consists mainly of mammals weighing from two hundred grams to two kilograms. This includes voles *(Microtus Spp.)*, rats, mice, and even foxes, marmots, and young deer. They also prey on birds as large as a Goshawk *(Accipiter gentiles)*, and will also go for grouse, corvids, woodpeckers, etc. In coastal areas, lake- and river sides they will feed on ducks, geese, and seabirds. They are known to take birds in full flight, and will on occasion even plunge into the water to get hold of a fish. They will also occasionally feed on reptiles and amphibians when the opportunity arises. Their strategy is to hunt from an open perch, taking their prey completely by surprise.

17.2.1 Biblical References

The Hebrew word *och* in Is 13:21 unfortunately is found nowhere else in the Old Testament. The Hebrew seems to be onomatopoetic (a sound imitation) of the species' call.

17.3 Eurasian Scops-owl

It should be noted that the vernacular name "Common Scops-owl" may refer to either of three species in the Scops-owl genus *Otus*. They were formerly considered conspecific (belong to the same species), and are allopatric, meaning that only one species is found in any given place:

- In Europe and western Asia, the European / Eurasian Scops-owl (*Otus scops*).
- In southern Asia, the Oriental Scops-owl (*Otus sunia*).
- In Africa, the African Scops-owl (*Otus senegalensis*).

The European Scops-owl is a small owl (about twenty centimeters long), and is distinguished from the Little Owl (*Athene noctua*) by its conspicuous ear-tuffs. It is well known for its territorial call, a deep whistle, similar to the call of the Midwife Toad (*Alytes Spp.*). The call is repeated at intervals of two to three seconds, for hours on end, making it an iconic sound of the African night life for example. It is uttered by both sexes. This owl is found in a wide range of habitats containing open and semi-open broadleaved woodlands and forests.

Eurasian Scops-owl (*Otus scops*)
(*Photo: Kevin Lin*) [3]

---

[3] This image is a reproduction of the photograph taken by Kevin Lin and used with permission. Originally placed on www.flickr.com, the original photograph can be viewed at: http://www.flickr.com/photos/hiyashi/5058957184/.

The parks and gardens, orchards, and cultivation with groves, as is found on the kibbutzim in Israel are sure to be inhabited by a pair. It feeds mainly on insects and other invertebrates. It nests in any kind of cavity and readily accepts nest boxes. Normally three to four eggs are laid in the cavity and then incubated by the female for about twenty-five days. The young leave the nest about twenty-eight days after hatching, and are able to fly by another twenty-eight days. They are cared for by the parents for about another five to six weeks. The chicks will reach maturity in about one year. To date, the oldest ringed bird to be observed again was found to be six years old. They are mostly migratory making short distance migrations to Africa for example. Their numbers have reportedly declined in northern Israel during the 1950's as a result of pesticide use, but the population appears to have stabilized since the 1970's.

17.3.1 Biblical References

The Hebrew word *sa'ir* is problematic in that it has three possible translations in the pointed text and three more in an unpointed text. In Lev 17:7 and 2 Chr 11:15 the context dictates it to be translated as *"demon."* In 2 Kng 23:8 the text reads "the high places of the gates", but a footnote suggests that because of a minute change in the spelling it should read "the high places of the demon/hairy being/Scops-owl." In my opinion the context would seem to favor the first option as a natural equivalent. I would agree with Koehier & Baumgartner that in Is 13:21 and 34:14 it could be the Eurasian Scops-owl that is meant.

17.4 Pallid Scops-owl

The Pallid Scops-owl *(Otus brucei),* is also known as the Striated Scops-owl, Bruce's Scops-owl, and the Desert Screech-owl. The Pallid Scops-owl is a small eared owl similar in appearance to the Common Scops-owl. However, it has more distinct streaks (striated) on the back and less intricate markings. It has a patchy distribution from Uzbekistan in the north to Pakistan in the south. It is also found in the southeastern corner of the Arabian Peninsula, and also westwards as far as Iran. It is rare in southern Israel and the adjoining parts of the Sinai Peninsula. It inhabits semi-open (almost Savannah-like) country with scattered trees and bushes. It's primarily an insectivore, but its diet is known to include things like lizards, spiders, and also small mammals. It occasionally hunts during the day, and has been known to take small bats and insects on the wing.

17.4.1 Biblical References

As the Hebrew word *tahmas* only occurs in the lists of "unclean" birds, namely in Lev 11:16 and Deut 14:15 it is not possible to ascribe it beyond any reasonable doubt to a specific species.

17.5 Little Owl

There are no less than thirteen recognized races of Little Owl *(Athene noctua)* spread across Europe and Asia. As its English name suggests, it is a small owl, about twenty-five centimeters in length. The Little Owl is a semi-desert dweller. The subspecies *Athene noctua lilith* is common in the Middle East. Stories about whom and what Lilith was are plentiful. She was purportedly a female demon of the night that supposedly flied around searching for newborn children either to kidnap or strangle them. Alan G. Hefner in his article in *Encyclopedia Mythica* states that one story is that God created Adam and Lilith as twins joined together at the back. She demanded equality with him, refusing to lie beneath him during sexual intercourse. She left him in anger and hurried to her home at the Red Sea, became a lover to demons and produced a hundred babies per day. Adam complained to God who then created Eve from Adam's rib, making her submissive to male dominance. In another version Adam supposedly had become tired of coupling with animals (a common Middle-Eastern herdsman practice, (declared a sin in Deut 27:21), and married Lilith. Refer to the Wikipedia comprehensive article on "Lilith"[5] for some interesting reading.

The Little Owl feeds mainly on small mammals and birds, reptiles, beetles, crickets and earthworms. Hunting usually takes place from dusk to midnight and then again just before dawn. It hunts by dropping on prey from a perch or running rapidly when chasing prey on

Little Owl *(Athene noctua)*
(Photo: Arturo) [4]

---

[4] This image is a reproduction of the photograph taken by Arturo, from Galicia (NW coast of Spain) and used with permission. Originally placed on www.flickr.com, the original photograph can be viewed at: http://www.flickr.com/photos/chausinho/2395948461/in/set-72157600950852011.

[5] Wikipedia contributors. "Lilith." Wikipedia, The Free Encyclopedia. Wikipedia, The Free Encyclopedia, 31 Jul. 2011. Web. 6 Aug. 2011.

the ground. Breeding is much the same as for the Common Scops-owl, but the Little Owl is essentially resident with chicks settling within twenty kilometers of their natal (birth) site.

17.5.1 Biblical References

The subspecies *Athene noctua saharae* is generally associated with the Hebrew word *kos* found in Lev 11:17, Deut 14:16, and Ps 102:7, and the subspecies *Athene noctua lilith*, with the Hebrew word *qa'ath* in Lev 11:18, Deut 14:17, Ps 102:7, Is 34:11, and Zeph 2:14. As indicated by its species name, the Blue Rock Thrush *(Monticola solitarius)* is a shy and unobtrusive bird, and as Douglas (p. 156) justly points out, could easily fit into many of the above contexts.

17.6 Northern Long-eared Owl

The (Northern) Long-eared Owl *(Asio otus)*,

Long-eared Owl *(Asio otus)*
(Photo: Mindaugas Urbonas) [6]

previously named *Strix otus*, is a medium-sized owl widely spread all over Europe and Asia. It is often associated with undulating, hilly, cultivated country areas, but when it comes to hunting it prefers open meadows, moors or marshland. It feeds mainly on small mammals like Voles *(Microtus Spp.)*, killing them by biting into the back of their sculls. Often birds are important food sources. They are active search hunters below the canopy in open forest areas, but like to utilize a perch especially in windy conditions. It locates most of its prey by audio clues, and can catch prey in complete darkness.

They use the old stick nests in trees that were built by other birds such as

---

[6] This image is a file from the Wikimedia Commons. Full details of the file can be found at: http://en.wikipedia.org/wiki/File:Long-eared_Owl-Mindaugas_Urbonas-1.jpg. Photo placed by Mindaugas Urbonas. This file is licensed under the Creative Commons Attribution-Share Alike 2.5 Generic license.

crows, ravens and magpies and various hawks. They will also readily use artificial nesting baskets when provided. Depending on the abundance of food, five to seven eggs are laid and incubated by the female for about twenty-seven days. During this time she takes short breaks at night and is fed by the male throughout. The chicks, although still flightless, leave the nest after three weeks to reside in the surrounding vegetation. They are capable of flying after five weeks, and reach maturity in about one year. An unusual characteristic of this species is its communal roosting in thickets during the winter months. The longevity record in the wild is twenty-seven years and nine months. They are mostly resident, but birds breeding in Europe may winter as far south as Egypt.

17.6.1 Biblical References

The Hebrew *janshoph* is found in Lev 11:17, Deut 14:16, and Is 34:11, probably referring to this species.

17.7 Short-eared Owl

The Short-eared Owl *(Asio flammeus)*, (*flammeus* is Latin for "flaming", or "the color of fire"), is a medium-sized owl about thirty-five centimeters in length. It raises its short "mouse ears" only when it is on guard or in a defensive mode. What makes it more conspicuous than other owls is the fact that it is more diurnal (active during the daytime rather than at night) than other owls. It hunts mostly in the late evening and early morning.

Short-eared Owl *(Asio flammeus)*
(Photo: Dario Sanches) [7]

It feeds mainly on small mammals including Voles *(Microtus Spp.),* Shrews *(Sorex Spp.),* Moles *(Scapanus Spp.),* Rabbits *(Oryctolagus Spp.),* and Hares *(Lepus Spp.).* It is distributed globally, and there are ten rec-

---

[7] This image is a reproduction of the photograph taken by Dario Sanches, from Sao Paulo, Brazil and used with permission. Originally placed on www.flickr.com, the original photograph can be viewed at: http://www.flickr.com/photos/dariosanches/5346439887/.

ognized subspecies. For example the Pueo or Hawaiian Short-eared Owl with the interesting scientific name of *Asio flammeus sandwichensis.* The Short-eared Owl prefers open country like marshes, grasslands, savannah and moorland. An almost unique behavior pattern amongst this family is that the female Short-eared Owl tends to build the nest. Breeding is much the same as the Long-eared Owl.

17.7.1 Biblical References

In Is 14:23 and 34:11 the Hebrew word *qippod* is also translated as "hedgehog." The fact that the other three creatures named by the author in Is 34:11 are birds, would seem to suggest that in this case *qippod* should rather be translated as "Short-eared Owl." The context in Zeph 2:14 confirms this, as a Hedgehog would not easily be found to "roost on her columns" (NIV).[8]

17.8 Non-biblical Species

From Chapter 6 it should be clear that our efforts to identify specific species in the Old Testament are often totally impossible. With the *Strigidae* it is very much the case. Although we have made choices in the texts mentioned above, we should keep in mind that at least three other members of this family do occur in the Middle East as well. Because of the close similarity of some of these to one of the above species, we have to take note of them, and admit that could in actual fact be one of them that the biblical author had in mind. These species are:

- Brown Fish-owl *(Ketupa zeylonensis)* feeds mainly on fish, frogs and freshwater crabs. It is therefore almost always found nearby water of some sort. The use of rodenticides and the draining of Lake Hula and the adjacent swamps in the 1950's have been implicated in the reduction in numbers of the species in Israel.
- Tawny Owl *(Strix aluco)* is a crow sized, plump owl with all-black eyes. It is mainly nocturnal, hunting between dusk and dawn. It likes to roost on a branch, pressed tight against the tree trunk, but its presence is often given away by the alarm calls of small birds, especially the Tits *(Parus Spp.).*

■ Hume's Owl *(Strix butleri)*, also known as Hume's Wood-owl or Hume's Tawny-owl, is a desert species found only in and around the Arabian Peninsula. In this area it is found in palm groves and near *Acacia* trees. In the late 1980's the Israel population was estimated at as low as two hundred pairs. Deaths because of road traffic are the main cause of mortality.

White fronted Bee-eater *(Merops bullockoides)*
(Photo: Rus Koorts) [9]

---

[9] This image is a reproduction of the photograph taken by Rus Koorts from Pretoria, in South Africa and used with permission. Originally placed on www.flickr.com, the original photograph can be viewed at: http://www.flickr.com/photos/ruslou/5742216665/in/set-72157625798426776

CHAPTER 18
ORDER: COLUMBIFORMES

> "Then he sent out a dove to see if the water had receded from the surface of the ground. But the dove could find no place to set its feet because there was water over all the surface of the earth; so it returned to Noah in the ark. He reached out his hand and took the dove and brought it back to himself in the ark."
>
> **Genesis 8:8–9 (NIV)**[1]

18.1 The Order

Columbiformes are an avian order that includes the very widespread and successful Doves and Pigeons, classified in the family *Columbidae*. The order includes the extinct Dodo and the Rodrigues Solitaire, long classified as a second family *Raphidae*, but lately considered as a subfamily *Raphinae* in the family *Columbidae*. A total of three hundred and thirteen species comprise the order *Columbiformes*, and they are found all over the world excluding the Arctic regions.

All *Columbiformes* species are monogamous, as is found with many other bird species. Unlike most other birds, however, they are capable of sucking up water, and can thus drink without having to tilt their heads back. Largely due to this drinking behavior, the Sandgrouse (*Pteroclidae*) were formerly included in this order.

These birds are normally medium sized, but some rather small and some rather large species do occur. For example the Blue-crowned Pigeon *(Goura cristata)* weighs about two kilograms, whereas the Common Ground-dove *(Columbigallina passerina)* weighs only about thirty grams. This, together with the fact that they are often easy to get hold of by humans, make them an important and highly sought after source of protein for humans in many parts of the world. In Brazil for example, one person may catch eight hundred birds per night by "spotlighting" them at their drinking holes after dark. (See par. 4.5).

Their ability to find their way home over extremely long distances, and their speed of flight resulted in the creation of the sport of pigeon racing. They are known to fly up to

---

a thousand kilometers per day, with one of the distance records 7,588 kilometers covered from England to Brazil by a bird called "Charlie." Recently a racing homer (homing pigeon) was sold for US$350,000. It is interesting to note that pigeons played a prominent role in Darwin's research and publications. He saw domestication as a process similar to speciation (the evolutionary development of a biological species) in which certain traits are selected or eliminated by nature in an isolated population. In this case he noted that all the known breeds of domestic pigeons descended from a single ancestor, the Rock Pigeon *(Columba livia)*.

Emerald-spotted Wood-Dove *(Turtur chalcospilos)*
(Photo: Rus Koorts) [2]

Special note on *Raphidae*: By studying sources like ships' logs and travelers' accounts of their trips all over the world, scientists and researchers have now determined without a doubt that this pigeon-like family consisted of three species, namely: The Dodo *(Ruphus cucullatus)*, the Rodrigues Solitaire *(Pezophaps solitaria)*, and the Reunion Solitaire *(Ornithaptera solitaria)*, all of whom are extinct today. (See par. 4.5). The Dodo has now become one of the most widely known symbols of extinction. Today, all that remains of the species are two feet, two heads and a number of skeletons kept in European museums.

---

[2] This image is a reproduction of the photograph taken by Rus Koorts, from Pretoria, South Africa and used with permission. Originally placed on www.flickr.com, the original photograph can be viewed at: http://www.flickr.com/photos/ruslou/3680260703/sizes/o/in/photostream/.

The Dodo is known to many people through Sir John Tenniel's illustrations for Lewis Caroll's *Alice in Wonderland*. He based his illustrations on those of various Dutch artists who actually saw the birds alive as they were brought to Europe by sailors in the sixteenth and seventeenth centuries.

It should also be noted that many other species in this order, although not officially crossed off the list, may well be extinct as they have not been sighted for long periods of time. Fifty-eight of the three hundred and nine species of Pigeons are currently listed as "Threatened", forty-seven of these are insular species (dwelling or situated on an island).

The subfamily *Columbinae* may be grouped together as the seed-eating species that are almost exclusively granivorous, as opposed to the fruit-eating Fruit Doves *(Ptilinopus)* who are mainly frugivorous.

CHAPTER 19
FAMILY: COLUMBIDAE

> "As soon as Jesus was baptized, he went up out of the water.
> At that moment heaven was opened, and he saw
> the Spirit of God descending like a dove and alighting on him."
>
> **Mat 3:16 (NIV)**[1]

19.1 Pigeons and Doves

This family could be called the typical Pigeons and Doves, and contain one hundred and eighty-one species. Laymen tend to call the smaller members of the family "doves", and the larger members "pigeons", but size alone does not always distinguish the two groups. For example: the Rock Dove *(Columba livia)*, is often called the "Rock Pigeon." They are characterized by the following features:

- Their bills are usually short, soft at the base, swollen at the tip, with a waxy cere (membrane) over the nostrils.
- They have short legs, and their toes are well adapted for perching and walking, as they are all terrestrial (land birds), as opposed to arboreal species, which live primarily in trees.

Rock Dove / Rock Pigeon *(Columba livia)*
(Photo: Rus Koorts) [2]

---

[1]   THE HOLY BIBLE, NEW INTERNATIONAL VERSION®, NIV® Copyright © 1973, 1978, 1984, 2011 by Biblica, Inc.™ Used by permission. All rights reserved worldwide.

[2]   This image is a reproduction of the photograph taken by Rus Koorts, from Pretoria, South Africa and used with permission. Originally placed on www.flickr.com, the original photograph can be viewed at: http://www.flickr.com/photos/ruslou/4931570546/.

- Their plumage is highly variable, but they are commonly known as grey birds with maybe iridescent patches especially on the neck and wings. These patches are like a soap bubble or butterfly wings which appear to change color as the angle of view or the angle of illumination changes. Being terrestrial birds, their upper parts would typically be cryptic (concealing), with maybe lighter under parts. This is an effective adaptation to camouflage them in their typical environment. The effectiveness of the Bare-faced Ground-dove *(Metriopelia ceciliae)* in this regard is remarkable, as it lives among the lichen-covered rocks of dry stony Puna country in the Andes. Sexual dichromatism (exhibiting two color phases within a species not due to age or season) is unusual amongst pigeons, but the Namaqua Dove *(Oena capensis),* is one such case. The fruit-eaters in contrast can be very gaudily colored (Latin: *gaudium,* meaning "enjoyment" or "merry-making", a striking feature that is distinguished by color and used to attract a female). In this case, generally bright green with ornaments of red, purple, pink, yellow, etc. The Crested Pigeon *(Ocyphaps lophotes)* of eastern Australia for example is unique in having a double ornamentation.

Namaqua Dove *(Oena capensis)*
(Photo: Rus Koorts) [3]

By using air accumulated in the crop, this family produces flute-like cooing noises (and sometimes whistles), with the bills kept closed. They are often gregarious, mostly arboreal (living in or among trees). They feed on plant material like seeds and fruit, and feed their young by regurgitating from a large crop. They need to drink water often so as to moisten their food, and to prepare "crop-milk." Their method of drinking is unique, as they are the only birds that are able to suck up water. They immerse their beaks and drink their fill without raising their heads, as practically all other birds have to do. They have thick heavy plumage,

---

[3] This image is a reproduction of the photograph taken by Rus Koorts, Pretoria, South Africa and used with permission. Originally placed on www.flickr.com, the original photograph can be viewed at: http://www.flickr.com/photos/ruslou/4628347267/sizes/o/in/photostream/.

with the feathers loosely attached in the skin. They are fond of sunning, the low intensity mode being when the bird lies on one side with the tail and the top wing spread out. Allopreening, the nibbling of one partner at the plumage of the other, occurs between potential or actual pairs. This might be one of the reasons why they are an icon of love in many cultures. Their flight is strong, with their wings making characteristic flapping sounds in flight.

Pigeons are monogamous, at least for a particular breeding season. The males use nest- and/or aerial displays to attract a female, with the bow-coo being the most well known strategy. They nest in suitable man-made structures, trees, on cliffs, or on the ground. The nest consists of a light platform of twigs that does not require a large amount of effort to build. They are unique in that they feed their chicks on a nutritious substance called "crop-milk." The nestlings are covered in a sparse down, which is soon replaced by the first feathers. They are distributed worldwide, except for the Polar Regions.

19.2 Non-Biblical References

Doves are often associated with fertility, probably because of their high reproductive capabilities. From the eastern Mediterranean comes a fertility figure from about 4500 BC, depicting a woman with a dove perched on her upraised arms. Although not all doves are peaceful, the goddesses of love from Rome (Venus) and ancient Greece (Aphrodite) are both associated with the dove. Kamadeva, the Hindu god of love, is depicted riding on a dove. See the next paragraph for a discussion of this theme in biblical Hebrew.

Rock Doves/Pigeons have been domesticated since antiquity, making it the first known domesticated avian species. They are used widely as food and also for the carrying of messages in military operations by for example the Assyrians, Egyptians, and Persians. The first documented record of the use of pigeons was by Prontinus, reporting that Julius Caesar used pigeons as such during a siege. During the Franco-Prussian war and both the World Wars there are records of pigeons saving many lives on the side of their masters.

19.3 Biblical References

The Rock Dove *(Columba livia)*, the Collared Dove *(Streptopelia decaocto)*, the Turtle Dove *(Streptopelia turtur)*, and the Laughing Dove *(Streptopelia senegalensis)*, are all common birds in the Middle East, often known as the Palm Dove. The Common

Laughing Dove *(Streptopelia senegalensis)*
(Photo: Sandeep Thoppil) [4]

Woodpigeon *(Columba palumbus)* is a rare wintering species, and the Namaqua Dove *(Oena capensis)* may occur accidentally from Arabia.

The Hebrew word *jonah* is a generic term usually translated as "dove." It is derived from the stem *anah* which means "to mourn", probably applied to the dove because of the call of some species like the Laughing Dove, for example. As in neighboring cultures (See par. 19.2), the Hebrew also has a love connotation, the noun being used as a term of endearment for a beloved girl on three occasions in the Song of Songs namely in 2:14, 5:2 and 6:9. In the sacrificial passages of Lev 5:7, 5:11, 12:6, 12:8, 14:22, 14:30, 15:14, 15:29, and Numbers 6:10 it is usually translated into English as a "young pigeon."

The Laughing Dove is also common in the Middle East, and would certainly be acceptable as sacrifices. However, the Hebrew word *tor* is a sound imitating word, based on the call of the Turtle Dove *(Streptopelia turtur),* as is the case with its scientific name. It is referred to in all the sacrificial passages of Lev 5:7, 5:11, 12:6, 12:8, 14:22, 14:30, 15:14, 15:29, and Numbers 6:10, making the sacrificial prescription unmistakable.

One of the most famous birds in Bibles is found in Gen 8:8–12, where the story about Noah trying to determine to what extent the floodwaters have receded is told. He sent out a raven, which did not return. But because he wanted to confirm that land was actually bare, he sent out the terrestrial dove. The Hebrew word *jonah* used here is a generic term usually translated as "dove." The Rock Dove was later renamed the Rock Pigeon. Simply because the pigeons are known for their homing skills, would I suggest that it actually was a Rock Dove / Pigeon.

---

[4]   This is a reproduction of the photograph taken by Sandeep Thoppil and used with permission. Originally placed on www.flickr.com, the original photograph can be viewed at: http://www.flickr.com/photos/sandeepthoppil/4424446378/.

In the Christian religion, the Holy Spirit is symbolized by a dove, as seen in Mat 3:16, Mrk 1:10, and Luk 3:22. Luk 2:24 refers to the sacrificial prescriptions of the Old Testament.

Common Ground dove (*Columbina passerina*)
(*Photo: Alan Vernon*) [5]

---

[5] This image is a reproduction of the photograph taken by Alan Vernon and used with permission. Originally placed on www.flickr.com, the original photograph can be viewed at: http://www.flickr.com/photos/alanvernon/5493054666/.

CHAPTER 20
ORDER: PASSERIFORMES

> "Like a crane or a swallow, so did I chatter: I did mourn as a dove:"
> **Is 38:14 (KJV)**

20.1 The Perching Birds / Songbirds

The order *Passeriformes* (Latin: *passer* which means "sparrow") are by far the largest bird order in terms of numbers. There are more than five thousand one hundred species in total or sixty percent of all known living birds. It is the most complex order (one hundred and fourteen families), and the most highly developed order of birds. In the animal kingdom, they form one of the most diverse terrestrial vertebrate orders. The order has roughly twice as many species as the largest of the mammal orders, the *Rodentia*. As stated, it contains over one hundred and fourteen families, the second most of any order of vertebrates. The *Perciformes*, the bony fishes has the most, with one hundred and fifty-five families and seven thousand species. The Passerines, as they are commonly known, are distributed all over the globe, except for the Arctic regions, and reach their greatest diversity in the tropics. To the layman they are known as the "Perching Birds", or the "Songbirds", but these terms are neither exclusive nor correct in scientific terms. They are also called the "Passerine Birds", simply referring to the order. As the design of their feet would suggest, all of the species are land birds. They are small to medium in size, with the largest the Ravens *(Corvidae)* exceeding one and a half kilograms in weight, and about seventy centimeters long, and the Australian Lyrebirds *(Menuridae)*. The smallest passerine is the Short-tailed Pygmy-Tyrant *(Myiornis ecaudatus)*, at six and a half centimeters long and weighing about four grams. The Pygmy Tit *(Psaltria exilis)* from Indonesia is about eight and a half centimeters long.

> "A bird does not sing because it has an answer.
> It sings because it has a song."
> **Chinese Proverb**

The feet of the birds are a distinctive feature of this order. They have three toes pointing forward, and one pointing backwards, all on the same level, and are always unwebbed. Another distinct feature of these birds is that they have a vocal organ called a syrinx (voice box) at the base of the bird's trachea, with which they are able to produce sounds. This is unlike the vocal cords that mammals use to utter sounds. In the middle nineteenth century, a German anatomist, Muller, pointed out that Passerines have different numbers of muscles in the syrinx. The so-called Suboscines, are considered more primitive as they have four or fewer pairs of syrinx muscles, whereas the "true Oscines" (songbirds) have five to eight pairs of syrinx muscles.

Some of the more unique families based on behavioral diversity and some of the more spectacular families based on species diversity are:

- The Ovenbirds *(Furnariidae)* for their varied nesting habits. They are named after the intricate clay nests, shaped like an old-fashioned outdoor oven that a few species build. In the same vein, the Mudnest Builders *(Grallinidae)* should be mentioned.
- The Dippers *(Cinclidae)* who are the only truly aquatic passerines. Their habitat and mode of life are shared by no other bird.

Indian Pitta *(Pitta brachyura)*
(Photo: Nidhin G. Poothully) [1]

- The Bowerbirds *(Ptilonorhynchidae)* who build such elaborate edifices that it is hard to believe that they are the work of one relatively small bird.
- Just as famous, but for a completely different reason, are the Birds of Paradise *(Paradisaeidae)*.
- To a lesser extent The Pittas *(Pittidae)*, the Sunbirds *(Nectariniidae)*, the Icterids *(Icteridae)* and the Lyrebirds *(Menuridee)*. But in other families of this order, like the Waxbills *(Estrildinae)* for example, there are some truly spectacular species as well.

Southern Yellow-Billed Hornbill
(Tockus leucomelas)
(Photo: Rus Koorts) [2]

---

[2] This image is a reproduction of the photograph taken by Rus Koorts from Pretoria, in South Africa and used with permission. Originally placed on www.flickr.com, the original photograph can be viewed at: http://www.flickr.com/photos/ruslou/5847937764/in/set-72157625798426776

CHAPTER 21
SUBORDER: PASSERI (OSCINES)
FAMILY: HIRUNDINIDAE

> "Like a fluttering sparrow or a darting swallow,
> an undeserved curse does not come to rest."
> **Prv 26:2 (NIV)**[1]

21.1 Swallows

The Swallows (and Martins) are a group of birds that are well known and loved by people all over the world. They are almost cosmopolitan (embrace multicultural demographics) in their distribution. In other words, they can be found on all continents except Antarctica. They consist of seventy-nine living species in about twenty genera. They have the following characteristics:

- They are small birds (ten to twenty-five centimeters long).
- They have sleek and slender bodies and long pointed wings.
- They have twelve rectrices (tail feathers) and facial bristles, as opposed to the Swifts *(Apodidae)* that have ten rectrices and no bristles.
- Their legs are short and their feet small, but they can easily perch on twigs, branches, wires and cables of all sorts, and they also like to sit on rooftops. When seen on the ground, they are usually gathering mud for plastering a nest or gathering other nesting material. They are able to walk short distances, but it is with difficultly and in an awkward shuffling kind of way.
- Their bodies are usually darker above than below, and their plumage often has some metallic sheen to it.
- They are peculiarly adapted to an aerial life, for example: their tiny triangular bills conceal their wide gapes (the interior of the open mouth).

Their flight is not as fast as those of the swifts and usually more erratic, but nevertheless flowing and graceful. They are most gregarious (social) and nest alone or colonially, in close association with man. The nest is padded with feathers and soft grass in a chamber at the end of a self-made burrow in a bank or on level ground. They also build a mud nest (consisting of a half-cup or bowl-and-tunnel structure) under a suitable overhang (natural or man-made). Mud nests are built by both sexes by plastering little pallets of mud like bricks, strengthening it with grass or straw. They are extremely nest faithful, often returning to exactly the same nest year after year to the delight of bird lovers all over the world. In South Africa many a home owner will anxiously await the return of a resident pair of Greater Striped Swallows *(Hirundo cucullata* syn. *Cecropis cucullata)* or Lesser Striped Swallows *(Hirundo abyssinica* syn. *Cecropis abyssinica).* Some species like the Tree Swallow *(Iridoprocne bicolor)* are cavity nesters. They use either natural cavities or nests that were abandoned by woodpeckers. Swallows lay up to six eggs, which are incubated by the female. The conscientious male will meanwhile sleep in a suitable spot nearby or beside the nest, making for an adorable site. Incubation of the eggs takes about fifteen days, depending on the attentiveness of the female. The chicks are sparsely downed when hatched, and cared for by both parents. The chicks are fully fledged after about twenty-two days. Swallows in general frequently produce a second brood and occasionally even a third per year, with the later clutches usually smaller than the first.

Especially after becoming involved in agriculture, humans have been looking for signs of the arrival of spring, and swallows have been a universal icon in this regard. They migrate by day, and need to feed while travelling, thus needing sufficient insects, which in turn are usually coupled to warmer weather. They therefore follow specific isotherms (lines connecting locations with equal temperatures), which explain the varying departure and arrival dates. The Common Swallow *(Hirundo rustica),* is the most familiar and widespread member of the family. It is known in Germany as the Chimney Swallow; in America, Norway and the Netherlands as the Barn Swallow; and in Southern Africa as the European Swallow.

21.2 Biblical References

In the Middle East region we find the Common Swallow *(Hirundo rustica)*, the Red-rumped Swallow *(Hirundo daurica)*, the Crag Martin *(Ptyonoprogne rupestris)*, the Pale Crag Martin *(Hirundo obsolete)*, the Sand Martin *(Riparia riparia)* (mainly in Egypt), the Rock Martin *(Ptyonoprogne fuligula)*, and the House Martin *(Delichon urbica)*.

In the Old Testament the Hebrew word *dror* is translated with "swallow." In Ps 84:4 it is said that swallows were nesting in the Temple area. In Prv 26:2 the aerial lifestyle of the swallow is implied.

Common Swallow *(Hirundo rustica)*
(Photo: Malene Thyssen) [2]

---

[2] This image is a file from the Wikimedia Commons. Full details of the file can be found at: http://en.wikipedia.org/wiki/File:Landsvale.jpg. This file is licensed under the Creative Commons Attribution 2.5 Generic license by Wikipedia Commons user Malene. malene.thyssen73@gmail.com.

CHAPTER 22
FAMILY: CORVIDAE

> "It shall be that you will drink of the brook,
> and I have commanded the ravens to provide for you there."
>
> **1 Kings 17:4 (NASB)**[1]

22.1 Ravens and Crows

This family contains over one hundred and twenty species, including some of the world's most familiar birds including:

- Treepies.
- Oriental Magpies.
- Old World, New World, and Grey Jays.
- Stresemann's Bush Crow.
- Nutcrackers.
- Holarctic Magpies.
- True Crows (Crows, Ravens, Jackdaws, and Rooks).
- Azure-winged Magpie.
- Choughs.

Ravens, and more so Crows, possess remarkable personalities, which can rightly be described as assertive and full of self-confidence. They are also seen as brash, bold, and aggressive. In addition to this, they are highly active, and particularly noisy. They possess an extensive dietary range, meaning they will eat almost anything that they can swallow, animal or vegetable, dead or alive. This, coupled with their boldness and intelligence, has enabled them to adapt completely to a lifestyle co-existent with man. This is the case even

---

[1] Scripture quotations taken from the New American Standard Bible®, Copyright © 1960, 1962, 1963, 1968, 1971, 1972, 1973, 1975, 1977, 1995 by The Lockman Foundation. Used by permission. (www.Lockman.org).

where the habitat has radically been altered by the destructive activities of humans and the use of their machines. Therefore they are distributed almost world-wide. However, they are absent from Antarctica, New Zealand, southern South-America, and certain oceanic islands.

They have hoarse, croaking calls that carry far. In agricultural areas they are usually welcomed by farmers as they consume enormous numbers of harmful insects and larvae. They could, however, in the process cause considerable damage to sprouting crops like grain. Because they are well-known for their love of the eggs and small chicks of other birds, sentimental bird-lovers are not fond of them at all. They are widely credited with being the most "intelligent" of all birds, but this is based on blatantly anthropomorphic evidence (the attribution of human characteristics, or characteristics assumed to belong only to humans, to animals). This might be illustrated by their imitative skills. These skills are, however, not as well developed as those of the Mynahs *(Sturnidae)* and the Parrots *(Psittacidae)*. The Crows *(Corvinae),* however, are said to have some counting skills.

22.2 Raven

The Hebrew word 'orev is usually translated as "raven", probably meaning the Common Raven *(Corvus corax),* but the Hooded Crow *(Corvus corone cornix),* is also a common sighting in cities and other urban areas. The Carrion Crow *(Corvus corone corone),* and to a lesser extent, the Rook *(Corvus frugilegus),* are common in the Middle East, and would probably fit well into any of the biblical references. Except for the Jackdaw *(Corvus monedula),* Modern Hebrew uses 'orev as the generic name for all of these species. Jackdaws are generally more cautious of people, and have thus not fared so well in competition with mankind

Common Raven *(Corvus corax)*
(Photo: David Hofmann) [2]

---

as some of their smaller relatives. They are therefore now found mostly in the wilder, uninhabited parts of their distribution range.

The Raven is the largest species in the Order *Passeriformes*. It grows up to eighty centimeters in length, and is characterized by:

- A strong unnotched bill.
- Nostrils covered with forward-directed bristles.
- Moderately long legs, and large strong feet.

22.3 Biblical References

Because of their extensive dietary range, they are included in the lists of "unclean" birds in Lev 11:15 and Deut 14:14. In 1 Kng 17:4, and 17:6 ravens are said to supply Elijah with food. Ravens are more shy and vigilant than the crows. They prefer mountainous terrain or solitary wooded areas suitable for nesting, and would fit into the context of the story. In Gen 8:7 we find probably one of the most well known references to birds in the Old Testament, as the raven is the first to leave Noah's ark. (See par. 27.2). Because of the difference in personality, I would like to believe that it in actual fact was one of the crows that were sent. Both Job 38:41 and Ps 147:9 suggest that parents in this family have their hands full with ravenous chicks. Prv 30:17 refer to the practice by crows to peck at the eyes of a fresh carcass. So 5:11 takes note of the raven's plumage, and Is 34:11 of its preference to live in solitary areas.

CHAPTER 23
FAMILY: PLOCEIDAE

> "Are not two sparrows sold for a copper coin?
> And not one of them falls to the ground apart from your Father's will."
> **Mat 10:29 (KJV)**

23.1 Sparrows

As its vernacular name (a name in general use within a community) indicates, the House Sparrow *(Passer domesticus),* is the most abundant and successful urban dweller of all birds. Step off a plane at any airport, and you will probably be welcomed by the chattering local family. The House Sparrow originated in the Middle Eastern region. Originally, they were native to Europe, western Asia and northern Africa, but they have since followed European civilization all over the world. They have settled successfully after being introduced intentionally and/or accidentally too many parts of the world like for example: Southern Africa, South America, Australia, New Zealand, and Hawaii. They were

House Sparrow *(Passer domesticus)*
(Photo: Dario Sanches) [1]

---

[1]   This image is a reproduction of the photograph taken by Dario Sanches from Sao Paulo, Brazil and used with permission. Originally placed on www.flickr.com, the original photograph can be viewed at: http://www.flickr.com/photos/dariosanches/6179214578/in/photostream.

introduced into North America in 1852 in a Brooklyn (New York) cemetery, and subsequently spread over the entire continent. They have been, however, unable to survive in Greenland. In the Philippines they are unable to compete with the Eurasian Tree Sparrow *(Passer montanus),* which had already been introduced there from Asia and occupies the same niche in the ecosystem.

"The House Sparrow is part of the Sparrow genus *Passer*, which contains about twenty species, depending on the authority one follows. A large number of subspecies have been named, of which twelve were recognized in the *Handbook of the Birds of the World.* The House Sparrow has been described as a "chunky" bird, typically about sixteen centimeters long. It has a large rounded head, a short tail, and a stout bill. Its weight generally ranges from twenty-four to forty grams. The plumage of the House Sparrow is mostly different shades of grey and brown, with some variations in the twelve subspecies. They are very sociable, gregarious in all seasons when feeding, roosting communally, and often form flocks with other types of birds. In birding terms this is called a "Bird party." Their nests are usually grouped together in clumps, and they engage in a number of social activities, such as dust and water bathing, and "social singing", in which birds call together in bushes."[2]

In 1972, W.A. Hickey wrote a book in Afrikaans entitled: *Twee vir 'n stuiwer* (Two for a Farthing), referring to the words of Jesus in Mat 10:29–31 and Luk 12:6–7. In it Hickey uses a Sparrow community as metaphors for human behavior. At the time it was prescribed reading in government schools and so became part of Afrikaans folklore.

23.2 Biblical References

In the Middle East region one can readily find the House Sparrow, the Spanish Sparrow *(Passer hispaniolensis),* the Rock Sparrow *(Petronia petronia),* and the Dead Sea Sparrow *(Passer moabiticus)* in their respective distribution ranges. Ps 84:4 and Ps 102:8 refer to a bird sitting alone on the housetop, which hardly suggests a sociable House Sparrow, and could be referring to the Blue Rockthrush *(Monticola solitarius)* which is a shy, solitary bird, as indicted by its name, that perches on rock summits and even houses.

---

[2] Wikipedia contributors. "House Sparrow." Wikipedia, The Free Encyclopedia. Wikipedia, The Free Encyclopedia, 2 Aug. 2011. Web. 7 Aug. 2011.

In the New Testament the references in Mat 10:29,31 and Luk 12:6,7 are in accordance with the custom to this day to sell sparrows alive and/or cooked in street markets all over the world.

Cape Sparrow *(Passer melanurus)*
(Photo: Rus Koorts) [3]

---

[3] This image is a reproduction of the photograph taken by Rus Koorts, Pretoria and used with permission. Originally placed on www.flickr.com, the original photograph can be viewed at: http://www.flickr.com/photos/ruslou/3867378541/.

CHAPTER 24
ORDER: APODIFORMES

> "Like a crane or a swallow, so did I chatter:
> I did mourn as a dove: mine eyes fail with looking upward:
> O LORD, I am oppressed; undertake for me."
>
> **Is 38:14 (KJV)**

24.1 Swifts, Tree Swifts, and Hummingbirds

"Traditionally the bird order *Apodiformes* contained three living families: the Swifts *(Apodidae)*, the Tree Swifts *(Hemiprocnidae)*, and the Hummingbirds *(Trochilidae)*. In the Sibley-Ahlquist taxonomy, this order is raised to a superorder *Apodimorphae* in which Hummingbirds are separated as a new order, *Trochiliformes*. With nearly four hundred and fifty species identified to date, they are the most diverse order of birds after the passerines.

As their name ("footless" in Latin) suggests, their legs are small and have limited function aside from perching. The feet are covered with bare skin rather than the scales (scutes) that other birds have. Another shared characteristic is long wings with short, stout humerus bones. The evolution of these wing characteristics has provided the hummingbird with ideal wings for hovering."[1]

The family *Apodidae* consists of four tribes:

- Tribe *Cypseloidini* Swifts.
- Tribe *Collocalini* Swiftlets.
- Tribe *Chaeturini* Needletails.
- Tribe *Apodini* Typical Swifts.

---

[1] Wikipedia contributors. "Apodiformes." Wikipedia, The Free Encyclopedia. Wikipedia, The Free Encyclopedia, 7 Jun. 2011. Web. 12 Aug. 2011.

CHAPTER 25
TRIBE: APODINI

"Even the stork in the sky knows her appointed seasons, and the dove,
the swift and the thrush observe the time of their migration.
But my people do not know the requirements of the LORD."

Jer 8:7 (NIV)[1]

25.1 Typical Swifts

The tribe *Apodini*, the Typical Swifts, consists of five genuses:

- Genus *Aeronautes* (three species).
- Genus *Tachornis*   (three living species).
- Genus *Panyptila*   (two species).
- Genus *Cypsiurus*   (two species).
- Genus *Apus*   (about seventeen species).

Apus is from the Greek meaning "without foot", as the feet of these birds are generally speaking completely unsuitable for perching. Their tiny feet, however, belie their effectiveness for gripping onto vertical surfaces. With two toes spread medially (to the middle) and two spread laterally (to the side), the swifts share a characteristic with the Chameleons *(Chamaeleo)* and the Koala *(Phascolarctos cinereus),* all three being climbing vertebrates *par excellence.*

The haemoglobin (a conjugated protein that is very important in the transportation of oxygen to the tissues in the body) in the blood of swifts is sensitized for the optimum delivery of oxygen in conditions of low oxygen pressure. This enables them to effortlessly maintain a high-altitude lifestyle, including feeding and drinking water. They are even thought to copulate in flight. They are characterized by:

- A very characteristic shape. Their short forked tails and very long swept-back wings resemble a crescent or a boomerang.
- Short and weak beaks. As with the Nightjars *(Caprimulgidae)* the gape (the interior of the open mouth) is exceptionally large, facilitating the aerial capture of insectivorous prey.
- Coarse, bristle-like feathers, located immediately in front of the eyes, which protect the eyes in flight.

Swifts will readily travel far away from their breeding or roosting sites in search of sufficient food supplies. However, because it plays an important part in their breeding cycle, they are seldom found far from some form of open water. They are highly gregarious (social) throughout the year, often forming large mixed flocks with other swift and aerial feeding species such as swallows. To date, aerial roosting (to spend the night on the wing), at an elevation of one thousand to two thousand meters has been proven beyond doubt for the Common Swift *(Apus apus)* only. Swifts roost according to the availability of prey, and they may even form daytime roosts during unusually cold or damp weather. This is done presumably for them to have a rest and conserve their body heat. Swifts are very vocal, especially in the breeding season, with the members of a pair constantly calling in a duet.

All swifts are exclusively insectivorous, with the size of the prey that they would prefer determined largely by the gape size (the interior of the open mouth). This, together with aerial stratification (arrangement in distinct layers/strata), results in a reduction in competition for food, and allows a species to specialize in a single insect species. For example, the stomach of a White-collared Swift *(Streptoprocne zonaris)* was found to contain nothing else but eight hundred winged ants.

Swifts are usually monogamous and pair for very long periods. The

Common Swift *(Apus apus)*
(Photo: Peter Schoen) [2]

---

[2] This image is a reproduction of the photograph taken by Peter Schoen and used with permission. Originally placed on www.flickr.com, the original photograph can be viewed at: http://www.flickr.com/photos/peterschoen/5863939185.

bond of a pair will be renewed annually in the case of migrant species. Nest sites are mostly in dark and dry places with clear access to the nest entrance. Bridges, viaducts, and towers are becoming more popular as natural sites diminish. Some species are now using man-made structures exclusively. Others, like for example the African Palm-swifts *(Cypsiurus parvus),* have specialized nest requirements. It is necessary for their eggs to be glued onto the nest by means of their saliva.

Most swift species mate at or near the nest, but because of the observation of some special aerial displays, the possibility of mid-air copulation has been considered by some observers. However, conclusive proof is obviously not easily obtained and is therefore still questioned by many experts. Their breeding times, as well as their success rates when breeding, are influenced mostly by the prevailing weather patterns. The amount and strength of the wind and the subsequent availability of food (because insects can be blown away), will have a definite influence. Nesting duties, like the regular cleaning of the nest for example, are shared equally by both the male and the female. The chicks are naked when hatched, and the nestling period will range from forty to fifty-six days. The bolus of food (a portion of food ready for digestion) brought to the nest by the parents, is regurgitated directly into the throat of the chicks. The chicks soon grow relatively fat in order to survive in the case of lower feeding rates when the parents have to leave temporarily to escape inclement weather.

Migration takes place in flocks, at high altitudes, by day and by night, and is known to be extremely fast. For example, an Alpine Swift *(Tachymarptis melba)* covered one thousand six hundred and twenty kilometers in three days. Long distance movements are, however, not only confined to the annual migration process. Temperate-breeding species will undertake round trips of up to as much as two thousand kilometers to avoid adverse weather conditions.

Few bird families have such an enigmatic relationship with humans as the *Apodidae.* Many species are no longer using traditional sites, but utilizing previously unknown nest-sites, ready-made by humans. This often leads to secondary problems, which in turn have necessitated "Swift friendly" measures to be taken. In Amsterdam, for example, for some time now, it has been an offence to obstruct their nest entrances. Their value as natural pest-controllers is clearly immeasurable. Even so, humans have destroyed the habitats of even more and more species. Over-harvesting of nests has dramatically diminished the populations of South-East Asian Swiftlets like for example the Edible-nest Swiftlet (*Aerodramus fuciphagus*). The trading in bird-nests is an extremely lucrative industry. In 1995 Indonesia exported nests, mainly to Hong Kong, in the order of one billion US Dollars, representing no less than sixteen million nests. Tragically, in 1998, the *Bangkok Post* reporting on the local industry, denounced the enormous sums being made by the concessionaires and Thai government, while no benefits whatsoever were apparently passed on to the local people.

Other than the bird-nest trade, swifts are not otherwise threatened or exploited by humans on large scales. They are difficult to catch, and not much of a meal. Accidental deaths by colliding with aircraft and power lines occur occasionally. African Palm-swifts *(Cypsiurus parvus)* are reported to have died after drinking chlorinated water from swimming pools in South-Africa. On occasion they are netted for food, but it does not pose a major threat.

25.2 Biblical References

Four of the six "European" Swifts namely the Little swift *(Apus affinis)*, Common Swift *(Apus apus)*, the Pallid Swift *(Apus pallidus)*, and the Alpine Swift *(Tachymarptis melba)*, occur in modern day Palestine.

Swifts are called *sis* in both Biblical and Modern Hebrew, probably as a result of the "see-see" call of most swifts. They are mentioned twice in the Old Testament, namely in Is 38:14, mentioning the voice of the bird; and in Jer 8:7, referring to the migratory behavior of the bird.

CHAPTER 26
GENERAL AND UNRESOLVED

> "In that day men will throw away to the rodents and bats
> their idols of silver and idols of gold, which they made to worship."
> **Is 2:20 (NIV)**[1]

26.1 Bats

The Biblical authors were not primarily naturalists or ornithologists, but utilized the limited knowledge about nature known to them at the time of their inspiration and subsequent innovative writing. To them all flying creatures were birds, and would therefore include the Bats *(Chiroptera).* Aristotle on the other hand, understood the mammalian nature of Whales (order *Cetacea*) and Dolphins (suborder *Odontoceti*, the toothed whales).

Mammal, noun: any vertebrate of the class *Mammalia*, having the body more or less covered with hair, the mother nourishes the young with milk from the mammary glands, and the mother giving birth to fully alive young. A single very interesting exception in this class is the egg-laying monotremes (for example, the Platypus (*Ornithorhynchus anatinus*).

Bats are mammals with webbed forelimbs that have developed into wings. They are therefore the only mammals that are naturally capable of true and sustained flight. Even to this day most laymen consider bats as being abhorrent. They roost and hibernate communally in caves and suitable man made structures, where they either hang down from the ceiling or cluster together in hollows and crevices. They would therefore be easily accessible as food during times when severe shortages are experienced. Therefore, it was necessary to mention them as "unclean" in the Mosaic Law.

---

[1] THE HOLY BIBLE, NEW INTERNATIONAL VERSION®, NIV® Copyright © 1973, 1978, 1984, 2011 by Biblica, Inc.™ Used by permission. All rights reserved worldwide.

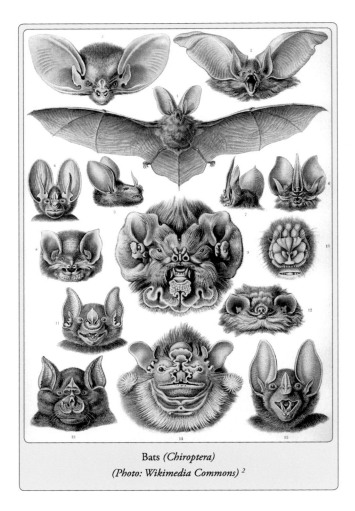

Bats *(Chiroptera)*
(Photo: Wikimedia Commons) [2]

26.1.1 Biblical References

Bats are insectivorous and for this reason included in the lists of "unclean" birds in Lev 11:19 and Deut 14:18, prohibiting Hebrews to turn to them for food. They are mentioned by the Hebrew name *'atalleph*. In Lev 11:20 "fowls that creep" are mentioned, and if one considers how a bat progresses on all fours when moving on a flat surface, this would seem to be a reasonable translation within this context. In Is 2:20 the abhorrence to bats by man and God is confirmed.

---

[2] This image is a file from the Wikimedia Commons. Full details of the file can be found at: http://upload.wikimedia.org/wikipedia/commons/f/fd/Haeckel_Chiroptera.jpg. This work is in the public domain in the United States because it is a work of the United States Federal Government under the terms of Title 17, Chapter 1, Section 105 of the US Code.

26.2 Unidentified Birds

26.2.1 Single Old Testament Occurrences (See par. 38.1)

The following Biblical Birds are mentioned only once in the Bibles. Because of the lack of any further cross references, we do not have conclusive evidence as to what bird, or possibly animal, is meant:

Griffon Vulture *(Gyps fulvus)*
(Photo: Ingrid Taylar) [3]

- In Is 34:15 the word *qippoz* is said to refer to a small serpent *(Coluber jugularis)* that lives in trees. But two owl species have also been suggested by Koehler & Baumgartner namely the Tawny Owl *(Strix aluco)*, and European Scops-owl *(Scops giu)*. The latter's name has been changed to Otus scops since the publication of their lexicon.
- The Hebrew word *barbur* is said to be a sound-imitating word, supposedly referring to the Lark-heeled Cuckoo, *(Centropus aegyptius)* in 1 Kng 5:3. Most translators have opted for words in the fowl-like family *(Phasianidae)*, but to this day the cuckoo is regarded a dainty morsel in Italy and Greece.
- Job 38:36 contains a word that could mean inward parts, but is also translated with "ibis."
- Eurasian Eagle-owl *(Bubo bubo)*, in Is 13:21.
- Griffon Vulture *(Gyps fulvus)*, in Mi 1:16.

26.2.2 Double Old Testament Occurrences (See par. 38.2)

The following Biblical Birds are mentioned only twice in the Bibles. Because of the lack of any further cross references, we do not have conclusive evidence as to what bird, or possibly animal, is meant. The first group is found only in the two listings of the "unclean" birds, and the other six are mentioned in texts spread over the Old Testament.

---

[3] This image is a reproduction of the photograph taken by Ingrid Taylar, Seattle, WA, USA and used with permission. Originally placed on www.flickr.com, the original photograph is seen at: http://www.flickr.com/photos/taylar/2609593617/.

- Cormorant *(Phalacrocorax carbo)*   Lev 11:17 and Deut 14:17.
- Pallid Scops-owl *(Otus brucei)*   Lev 11:16 and Deut 14:15.
- Hoopoe *(Upupa epops)*   Lev 11:19 and Deut14:18.
- Red Kite *(Milvus milvus milvus)*   Lev 11:14 and Deut 14:13.
- Egyptian Vulture *(Neophron percnopterus)*   Lev 11:18 and Deut 14:17.
- Lappet-faced Vulture *(Torgos tracheliotus negevensis)*   Lev 11:13 and Deut 14:12.
- Sea-gulls are called *shahaph* in Lev 11:16 and Deut 14:15. No less than eleven species of the family *Laridae* occur in the Middle East. Two of the more common ones are the Black-headed Gull *(Larus ridibundus)*, and the Little Gull *(Larus minutes)*.
- The Hebrew word *'anaphah* only occurs in the list of "unclean" birds in Lev 11:19 and Deut 14:18. Modern Hebrew utilizes this word as a generic term for the Herons *(Ardeidae)* found in the Middle East, namely the Grey Heron *(Ardea cinerea)*, Purple Heron *(Ardea purpurea)*, and the Goliath Heron *(Ardea goliath)*. The Biblical Hebrew stem from which this word is derived means "to be angry." Maybe the overall impression left by looking at the face of a heron might have led to this deduction.
- Common Barn-owl *(Tyto alba)* is *tinshemet*, in Lev 11:18 and Deut 14:16.
- Bearded Vulture *(Gypaetus barbatus)* is *peres*, in Lev 11:13, and Deut 14:12.
- The Hebrew word *dajjah* is found in Deut 14:13 and Is 34:15. The choice of the Griffon Vulture *(Gyps fulvus)* is based solely on the fact that the species is so well known today. Modern Hebrew actually calls the Griffon Vulture *nesher* which is known in Biblical Hebrew as an eagle or a vulture.
- Indian Peafowl *(Pavo cristatus)*, Hebrew: *tukkijjim*, in 1 Kng 10:22 and 2 Chr 9:21.
- Swallow *(Hirundinidae)*, Hebrew: *dror* in Ps 84:4, and Prv 26:2.
- Sand Partridge *(Ammoperdrix hayi)*, Hebrew: *qore'* in 1 Sam 26:20 and Jer 17:11.

Hoopoe *(Upupa epops)*
(Photo: Arturo) [4]

---

[4]   This image is a reproduction of the photograph taken by Arturo, from Galicia (NW coast of Spain) and used with permission. Originally placed on www.flickr.com, the original photograph can be viewed at: http://www.flickr.com/photos/chausinho/2396755210/.

- Yellow-vented Bulbul *(Pycnonotus xanthopygos)*, Hebrew: *'agur* in Is 38:14 and Jer 8:7.
- Swift *(Apus)*, Hebrew: *sis*, in Is 38:14, and Jer 8:7.

26.3 Unresolved Texts

26.3.1 Isaiah 34:11

The four words in the first sentence of this verse apparently refer to birds or animals. The word *qa'ath* appears in the lists of "unclean" birds in Lev. and Deut., removing any doubt that a bird is meant. Smith (p. 496) is of the opinion that a Pelican *(Pelecanidae)* resting its bill on its breast portrays a melancholy mood, prompting earlier translators in that direction. The only other occurrence of this word is found in Ps 102:7, where it is said to be in the desert. Although Pelicans are usually associated with wide open waters, they are on occasion found in dry areas. While traveling in northwestern Botswana, I have come across an adult Great White Pelican *(Pelicanus onocrotalus)*, perched on (or stuck in?) the top of a large *Acacia* tree. In Zeph 2:14 the first two names *qa'ath* and *qippod*, are used in

Great White Pelican *(Pelicanus onocrotalus)*
(Photo: Gay Biddlecombe) [5]

---

[5] This image is a reproduction of the photograph taken by Gay Biddlecombe LRPS, website: http://amberlight1.com and used with permission. Originally placed on www.flickr.com, as "amberlight1", the original photograph can be viewed at: http://www.flickr.com/photos/28119819@N07/5457120612/in/photostream/.

Southern Ground-Hornbill *(Bucorvus leadbeateri)*
(Photo: Rus Koorts) [6]

the same sequence as here. The only other occurrence of *qippod* is in Is 14:21. The only other occurrence of the corresponding stem is in Is 38:2, were it is translated as: "to roll up", as a weaver would his fabric. This might have prompted earlier translators to think in terms of the Hedgehog *(Erinaceous auristus* or *Erinaceous sacer)* or a Porcupine *(Hystricidae)*. In Zeph 2:14 the creatures are said to "spend the night on the capitals of pillars", excluding the Hedgehog in my mind. Douglas mentions that scholars have suggested the Bittern *(Botaurus stellaris),* which is common in the Middle East, as well as a Night Heron *(Nycticorax nycticorax)* and a Great Bustard (*Otis tarda*), all known for their strange booming calls at night, attributing to the idea of desolation.

Koehler & Baumgartner list two further possibilities namely the Little Owl *(Athene noctua lilith)* for the former, and the Short-eared Owl *(Asio flammeus),* for the latter. Both species are common in the Middle East, and known to inhabit solitary places.

The third term *yanshoph* posses an even greater problem as it occurs only here and in the lists of "unclean" birds. Of interest to note is that Modem Hebrew have utilized this word as the generic indication of the previous two species, calling them literally "owl ears" and "owl fields." Koehler & Baumgartner name the Long-eared Owl *(Asio otus)* as a possibility, which, in the absence of any further clues, I would go along with. The word 'orev poses no real problem, and so the more recent translations consider them to be "three owls and a raven."

---

[6] This image is a reproduction of the photograph taken by Rus Koorts from Pretoria, in South Africa and used with permission. Originally placed on www.flickr.com, the original photograph can be viewed at: http://www.flickr.com/photos/ruslou/5013541373/in/set-72157624830170739

26.3.2 Job 40:24–29

Six species of the Plover family *(Charadriidae)* occur in the Middle East, but it is the Spurwinged Plover *(Vanellus spinosus)* that is the most common.

From the context in which Job 40:29 is found, it would seem that the bird that is meant here is the Egyptian Plover *(Pluvianus aegyptius)* who is known to extract lice and other vermin from the jaws of an obliging crocodile. Thus the German name: *Krokodilvogel*.

26.3.3 Jeremiah 8:7

Some experts suggest that this first bird maybe a reference to the Crane *(Gruidae)* family, because of their migratory behavior. In the other references where this Hebrew word *hasidhah* is found, namely in Job 39:13, Zech 5:9, and Ps 104:17, migration does not feature. In the Middle East the Storks *(Ciconiidae),* are well known because of this annual phenomenon, with the White Stork *(Ciconia ciconia)* the most spectacular in numbers. Modern Hebrew has opted to use the Biblical name when referring to the latter. I would therefore go along with that interpretation, simply because of the lack of any tangible evidence to the contrary.

White-bellied Sunbird *(Cinnyris talatala)*
(Photo: Rus Koorts) [7]

---

[7] This image is a reproduction of the photograph taken by Rus Koorts from Pretoria, in South Africa and used with permission. Originally placed on www.flickr.com, the original photograph can be viewed at: http://www.flickr.com/photos/ruslou/4821493124/in/set-72157623311955116

26.3.4 Ezekiel 13:20

In this text there are two problematic verbs. Combined they have led to a wide range of translations varying in their degree of paraphrasing. If read in conjunction with Lam 3:52, the first verb means; "to hunt birds." The second is usually translated as "sprout, blossom or bud", but this particular conjugation is also connected to "fly." Literally the text then says: "Therefore, so said the Lord Yahweh: Behold, against the bands of charm that you hunt with them the souls that bud/sprout/fly." Taking the above, the macro-context, and the footnotes in the Masoretic Texts into consideration, this text then does not mention birds directly or even indirectly. An idiomatic translation could for example read: "Therefore, so said the Lord Yahweh: Behold, (I am) against the bands of charm with which you hunt (spiritually) budding souls."

Chinspot Batis *(Batis molitor)*
(Photo: Rus Koorts) [8]

---

[8]   This image is a reproduction of the photograph taken by Rus Koorts from Pretoria, in South Africa and used with permission. Originally placed on www.flickr.com, the original photograph can be viewed at: http://www.flickr.com/photos/ruslou/3744691113/in/set-72157617813371374

BIRDS AND BIBLES
IN HISTORY

PART 4
CONCLUSIONS

"The Hebrew Canon, as expounded in the Mosaic Law, the Prophets, and the Writings, contains the origins of Ornithology, and the biblical authors concerned, and not Aristotle, were the original Ornithologists."

Indian/Blue Peafowl *(Pavo cristatus)*
(Photo: GollyGforce) [1]

CHAPTER 27
CONCLUSIONS

"Does the hawk take flight by your wisdom
and spread its wings toward the south?
Does the eagle soar at your command and build its nest on high?"

Job 39:26–27 (NIV)[2]

27.1 Concluding Statement

"The Hebrew Canon, as expounded in the Mosaic Law, the Prophets
and the Writings, contains the origin of Ornithology,
and the biblical authors concerned, and not Aristotle,
were the original Ornithologists."

In the light of, and as a result of the preceding chapters, I am absolutely convinced that the basic hypothesis stated above and at the outset of this book, is indeed a true reflection of the facts presently before us. The church has played a major role in the initial developments of Ornithology. The main contributing factor being that throughout the centuries before the Renaissance, the clergy had been the intellectual core of society. From this position they determined what people were allowed to read and thus know and believe. However, the text of the Old Testament itself, and not the church, provides us with some extraordinary evidence substantiating the above hypothesis. Two examples should suffice.

27.2 Genesis 6–8.

It is extremely dangerous to draw conclusions from Gen 1–11, but in the case of the "Great Flood" we have many confirmations from extra-biblical sources. The so-called

---

"flood myths" in many cultures suggest that something happened, capturing the imagination of ancient man. In the recent past there have been attempts to scientifically motivate the possibility of an earth-covering flood. I, for one, am not convinced, yet. However, I firmly believe that some natural phenomena did occur, but where, when and on what scale, I am not prepared to speculate. That somebody because of divine intervention believed it necessary to take evasive action, I am prepared to accept, even if this included humans taking care of other living creatures. The only indication to the extent of this incident becomes clear if we keep in mind that the concept had been transferred orally for many generations until eventually being recorded. The author of Genesis, for example, deemed it necessary to describe the unfolding of this particular drama in the first part of his book that he was writing. His objective was to describe the results when an individual acts in obedience to a divine command against all odds. To what extent the whole story had been the product of an overly productive imagination is of no consequence here. The fundamental premise involved is: when something had been told or written, somebody had to possess sufficient intellectual resources to produce such a narrative. These intellectual resources originate from observations of either natural or manipulated phenomena, which today are known as field observations and laboratory research. The following example is a case in point: Somebody, let us also call him Noah, reportedly took fourteen specimens (seven pairs, and not one, as is commonly envisaged) of each living species into a watertight vessel. Commonsense should prevail here and make us realize that containing all of the species of the world in a single vessel is a matter of impossibility. In addition, today the symbolic value of the number seven is widely accepted by laymen and scholars of all religious persuasions.

Notwithstanding this, it has to be admitted that here the author's intentions were twofold:

- To describe the first attempt by a human being to not only consider the survival of himself and his kind but to, at the same time, prevent the eminent extinction of all or a number of species in the face of detrimental circumstances.

- But even more important: to utilize the above incident to convey the message that obedience to a divine command, is in the best interest of man and beast. At that time, and in those circumstances it meant engaging in completely irrational behavior. In other words, acting in faith was the right thing to do.

The incident provides us with the first ever mention to specific zoological families, and it is no other than two bird families, namely the family *Columbidae*, known as the Pigeons and Doves, and the family *Corvidae,* known as the Ravens and Crows. In Chapter 22 it was said of the latter: "their boldness and sagacity, coupled with their extensive dietary range, have enabled them to adapt completely to a lifestyle co-existent with man, even where the latter has radically altered the habitat." A radical change in precipitation patterns in a specific location would result in the habitat being significantly altered, albeit temporarily. It is clear that the narrator was an Ornithologist to the extent that he realized that if there was one species that would immediately be able to adapt to the circumstances out there and survive, it would be *Corvinae*. He observed bird behavior to the extent that he knew that Doves and Pigeons *(Columbiformes)* are generally terrestrial birds, and as long as the dove returned to its perch it would be an indication that the ecosystem had not become suitable for human habitation yet. The Hebrew Canon containing these narratives had fully developed by the time of Aristotle, and the accuracy of the ornithological facts contained in this narrative and other Scriptures were never questioned by him or anyone else to this day.

27.3 Job 39–41

Chapters 39–41 of the book Job must, apart from its literary genius, be recognized as the first written recordings from the pen of a naturalist and ornithologist *par excellence*. His descriptions of the might of the Hippopotamus *(Hippopotamus amphibius)*, the unassailability of the Crocodile *(Crocodylidae),* and the splendour of the Ostrich *(Struthiones),* are accurate in the finest detail. Aristotle had written more widely and his publications contain more investigative material, than the book of Job. It should however be kept in mind that the book Job is *sui generis* (of its own peculiar kind) in the Hebrew literature, written in a most distinctive language, and containing a timeless theology. Notwithstanding the fact that the author had no intention of demonstrating his zoological knowledge to his audience, one cannot but conclude that the above-mentioned chapters are the result of meticulous and exceedingly accurate observations. They clearly demonstrate that whoever made these observations, had a working knowledge and understanding on a wide variety of zoological and ornithological phenomena, comparable to those of Aristotle.

The dating of Job to at least 600 to 400 BC is the result of research by scholars from a wide range of disciplines conducted over many years. It confirms the fact that ornithology had been practiced long before Aristotle. (See par. 1.2).

27.4 The Prophets

As apposed to the twelve so-called "Small prophets", Isaiah, Jeremiah, Ezekiel and Daniel, are known as the "Large prophets." Isaiah and Jeremiah have utilized birds extensively in their books. (See par. 32). In motivating the above-mentioned concluding statement, their dating is of importance to us. (Par. 1.2).

27.5 The Future

With the above in mind, I believe that it is justified to make the following statement to serve as a request to all those concerned: "Future ornithological publications, when referring to the origins of Ornithology and the original Ornithologists, should unequivocally credit the Hebrew Canon in general, and more specifically, the relevant authors, as such."

It is my humble opinion that by giving such an unbiased record of the facts, it would greatly enhance the credibility of any publication. I hope that this publication would act as a catalyst in prompting others, undoubtedly more able than me, to study the subject further, gain more information, and hopefully answer the outstanding questions more fully.

When studying the Biblical Birds, I have come under the impression that the more recent translations have been

House Crow (*Corvus splendens*)
(*Photo: Dr. Marcel Holyoak*) [3]

more successful in finding the most natural equivalent to the meaning of the original author. I do believe that the work done by ornithologists over the past years have contributed tremendously to bring us to our present level of understanding. It goes without saying that we should keep on exploring, and so create the possibility of some exciting discoveries.

---

[3] This image is a reproduction of the photograph taken by Dr. Marcel Holyoak and used with permission. Originally placed on www.flickr.com, the original photograph can be viewed at: http://www.flickr.com/photos/maholyoak/5917587575/.

CHAPTER 28
POSTSCRIPT

> "The first living creature was like a lion, the second was like an ox,
> the third had a face like a man, the fourth was like a flying eagle."
>
> **Rev 4:7 (NIV)**[1]

28.1 Postscript

The closing words are not by me, but from two authors many times more eloquent, knowledgeable and wise than me. The first is Nigel J. Collar, Status and Conservation Consultant of the *Handbook of the Birds of the World*, writing in 1999 AD, and the second is the author of the Old Testament book called Numbers, writing three thousand years before Collar. Collar, in his essay *Risk Indicators and Status Assessment in Birds*, in Vol. 5 of *Handbook of the Birds of the World* wrote:

"Birdwatchers and biologists get to even more remote places - islands, interiors - by virtue of new airports, new logging roads, and new tourist facilities. They arrive as tiny components of the great machinery of economic development, which, in a few short years, mutilates natural landscapes and human cultures beyond recognition and brings Coca-Cola, television, chainsaws, DDT and debt to every cultivable corner of the planet. By the year 2010 not only will we know more about birds than ever before; we will also have most of them *completely surrounded*. We can therefore anticipate learning very early in the new millennium the full meaning of the notion of global stewardship; if not, we will discover instead what it might be like to visit the zoo in which all of the cages are empty."

He ends his essay by saying: "Growth in the composition of Red Lists provides some evidence for the way in which the world is coming to an end - perhaps not a literal end, but the end of an era to be sure. The year 2000 is a ridge top from which the planet's lovers of wild places can look back at a rich landscape haunted by wasted opportunities, and

---

peer forward at a dustbowl signposted with wasted words, "biodiversity" prime among them. So if we want the decidedly crumbly future to be worth the walk, we have to set about the pursuit of information and, in direct consequence, the pursuit of conservation, with an altogether new and messianic intensity. There is no reason why this needs to be left to professionals; amateurs can (and do) make a huge difference." (del Hoyo, et al, Vol. 5, p. 25).[2] That means me and you too!

To all who share my love for the birds, I quote from Numbers 6:24–26 according to the NIV:

"May the Lord bless you, and keep you;
May the Lord make his face shine upon you, and be gracious to you;
May the Lord turn his face towards you, and give you peace."

---

2   del Hoyo, J., Elliot, A. & Sargatal, J. eds. (1999). Handbook of the Birds of the World. Vol. 5. Barn-Owls to Hummingbirds. Lynx Edicions, Barcelona. Used with permission.

CHAPTER 29
BIBLIOGRAPHY

"the horned owl, the screech owl, the gull, any kind of hawk."
Deut 14:15 (NIV)[1]

29.1 Bibles

Aland, K., et al, (Eds.). (1966). *The Greek New Testament*. Stuttgart: United Bible Societies.

Bible Society of South-Africa (1957). *Die Bybel in Afrikaans*. Naslaan Uitgawe. Suffolk: Richard Clay (The Chaucer Press) Ltd.

Bible Society of South-Africa. (1973). The Holy Bible. *New International Version*. Cape Town: National Book Printers.

Bible Society of South-Africa. (1977). Good News Bible. *Today's English Version*, British Usage. Cape Town: National Book Printers.

Bible Society of South-Africa. (1983). *Die Bybel in Afrikaans*. Cape Town: National Book Printers.

Bible Society of South-Africa. (1998). *Verwysingsbybel. 1983-Vertaling*. Cape Town: National Book Printers.

Coverdale House Publishers. (1975). *The Living Bible*. London: Hazell Watson & Viney Ltd.

Dake, F.J. (1963). *Dake's Annotated Reference Bible*. Lawrenceville: Dake Bible Sales, Inc.

Elliger, K., Rudolph, W. (1967/1977). *Biblia Hebraica Stuttgartensia*. Stuttgart: Deutche Bibelgesel lschaft.

Gemser, B., et al. (1958). *Die Bybel met Verklarende Aantekeninge*. Kaapstad: Verenigde Protestantse Uitgewers (Edms.) Bpk.

Jones, A. (Ed.). (1974). *The Jerusalem Bible, Popular Edition*. London: Darton, Long man & Todd.

29.2 Dictionaries

Bosman, D.B., et al. (1984). *Tweetalige Woordeboek. Bilingual Dictionary*. Kaapstad: Tafelberg Uitgewers.

Brown, F., et al. (1906). *The Brown-Driver-Briggs Hebrew and English Lexicon*. Massachusetts: Hendrickson Publishers, Inc.

Davidson, B. *The Analytical Hebrew and Chaldee Lexicon*. London: Samuel Bagster and Sons Ltd.

Douglas, J.D. (Ed.), (1962). *The New Bible Dictionary*. London: Inter-Varsity Press.

Holladay, W.L. (Ed.). (1988). *A Concise Hebrew and Aramaic Lexicon of the Old Testament*. Leiden: E.J. Bull.

Inbal, S. (1993–4). *English-Hebrew, Hebrew-English Pocket Dictionary*. Jerusalem: S. Zack & Co., Publishers.

Koehler L., Baumgartner, W. (1958). *Lexicon in Veteris Testamenti Libros*. Leiden: E.J. Brill.

Kritzinger, M.S.B., et al. (1981). *Groot Woordeboek. Afrikaans-Engels*. Cape Town: J.L. van Schaik (Pty)

Meine, F. J. (Ed.). (1946). *The Consolidated Webster Encyclopedic Dictionary*. Chicago: Consolidated Book Publishers.

Smith, W. (2000). *Smiths Bible Dictionary*. Peabody: Hendrickson Publishers, Inc.

Sykes, J.B. (Ed). (1982). *The Concise Oxford Dictionary*. Oxford: University Press.

29.3 Commentaries

Alexander, D & P. (1986). *Handboek by die Bybel*. Kaapstad: Verenigde Protestantse Uitgewers (Edms)

Guthrie, D., et al. (Eds.). (1970). *The New Bible Commentary Revised*. London: Intervarsity Press.

Westermann, C. (1966). *Isaiah 40–66*. London: SCM Press Ltd.

> "What kind of bird are you, if you can't fly?" said he.
> To this the duck replied, "What kind of bird are you if you can't swim?"
> and dived into the pond.
>
> **From: Peter and the Wolf (1936) by Sergei Prokofiev (1891–1953)**
> **(Russian Composer, Pianist and Conductor)**

29.4 Birds

Austin, O.L. Jr. (1983). *Birds of the World*. Optimum Books.

del Hoyo, J., et al. (Eds.). (1992). *Handbook of the Birds of the World. Vol. 1. Ostrich to Ducks*. Barcelona: Lynx Editions.

del Hoyo, J., et al. (Eds.). (1994). *Handbook of the Birds of the World. Vol. 2. New World Vultures to Guineafowl*. Barcelona: Lynx Edicions.

del Hoyo, J., et al. (Eds.). (1997). *Handbook of the Birds of the World. Vol. 4. Sandgrouse to Cuckoos*. Barcelona: Lynx Edicions.

del Hoyo, J., et al. (Eds.). (1999). *Handbook of the Birds of the World. Vol. 5. Barn-Owls to Hummingbirds*. Barcelona: Lynx Edicions.

Great White Egret *(Ardea alba)*
(Photo: Dario Sanches) [2]

Harrison, J.A., et al. (Eds.). (1997). *The Atlas of Southern African Birds. Vol. 1. Non-passerines*. Johannesburg: BirdLife South Africa.

Harrison, J.A., et al. (Eds.). (1997). *The Atlas of Southern African Birds. Vol. 2. Passerines*. Johannesburg: BirdLife South Africa.

Jonsson, L. (1999). *Birds of Europe, with North Africa and the Middle East*. London: Christopher Helm (Publishers) Ltd.

Lockwood, G. (2000). *Birding with Sappi and Geoff Lockwood*. Centurion: Rapid Commercial Print Brokers and Publishers.

Maclean, G.L. (1993). *Roberts' Birds of Southern Africa*. Cape Town: The Trustees of the John Voelker Bird Book Fund.

Shirihai, H., et al. (2000). *A Guide to the birding Hot-spots of Northern Israel*. Society for the Protection of Nature in Israel.

---

[2]   This image is a reproduction of the photograph taken by Dario Sanches, from Sao Paulo, Brazil and used with permission. Originally placed on www.flickr.com, the original photograph can be viewed at: http://www.flickr.com/photos/dariosanches/6167690949/

Common Gull (*Larus canus*)
(*Photo: Arturo*) [3]

Shirihai, H., et al. (2000). *A Guide to the birding Hot-spots of Southern Israel*. Society for the Protection of Nature in Israel.

Sinclair, I., Hockey, P., Tarboton, W. (1997). *Sasol Birds of Southern Africa*. Cape Town: Struik Publishers.

Tarboton, W., Erasmus, R. (1998). *Owls and Owling in Southern Africa*. Cape Town: Struik Publishers (Pty) Ltd.

Unknown, (1995). *Atlas of Birds of China*. Shenzhen: Henan Science and Technology Press.

29.5 Biblical Hebrew

Deist, F.E. (1988). *Witnesses to the Old Testament*. Pretoria: N.G. Kerk Boekhandel (Edms) Bpk.

Gemser, B. (1968). Hebreeuse Spraakkuns. Pretoria: J.L. van Schaik Bpk.

Lamdin, T.O. (1973). *Introduction to Biblical Hebrew*. London: Darton, Longman & Todd.

Van der Merwe, C.H.J., et al. (1997). *'n Bybels-Hebreeuse Naslaangrammatika*. Kaapstad: Nationale Boekdrukkerye-groep.

Waltke, B.K., O'Connor, M. (1990).*An Introduction to Biblical Hebrew Syntax*. Winona Lake: Eisenbrauns.

---

[3] This image is a reproduction of the photograph taken by Arturo, from Galicia (NW coast of Spain) and used with permission. Originally placed on www.flickr.com, the original photograph can be viewed at: http://www.flickr.com/photos/chausinho/2372311632/in/set-72157600950852011.

Weingreen, J. (1959). *A Practical Grammar for Classical Hebrew*. London: Oxford University Press.

29.6 General

Abbot-Smith, G. (1964). *A Manual Greek Lexicon of the New Testament*. Edinburgh: T & T Clark.

Boolootian, R.A., Stiles, K.A. (1981). *College Zoology*. (10th edition). New York: Mac-Millan Publishing Co., Inc.

Canning, J. (Ed.). (1965). *100 Great Modem Lives*. London: Odhams Books Limited.

Cinamon, D.V. (1984). *Die Nuwe Vertaling Bybelkonkordansie*. Ravenmoor: Sceptre Uitgewers.

Deist, F. (Ed.). (1982). *Die Bybel Leef*. Pretoria: J.L. van Schaik Bpk.

Fisher, H.A.L. (1936). *A History of Europe*. London: Edward Arnold & Co.

Gordon, S. (1999). *John Denver, The Ultimate Hit Collection*. RCA Victor.

Martinez, F.G. (1992). *The Dead Sea Scrolls Translated. The Qumran Texts in English*. Leiden: E.J. Brill/William B. Eerdmans: Grand Rapids.

Mellor, E.B. (Ed.). (1972). *The Making of the Old Testament*. Cambridge: University Press.

Mills, C., Hes, L. (Eds.). (1997). *The Complete Book of South-African Mammals*. Cape Town: Struik

Nida, E.A., Taber, C.R. (1982). *The Theory and Practice of Translation*. Leiden: E.J. Bull.

Rousseau, L. (1974). *Die Groot Verlange. Die verhaal van Eugene N. Marais*. Cape Town: Human & Rousseau Uitgewers (Edms.) Bpk.

Rousseau, L. (1982). *The Dark Stream. The story of Eugene Marais*. Johannesburg: Jonathan Ball.

Rowley, H.H. (1960). *The Teach Yourself Bible Atlas*. London: The English Universities Press Ltd.

Ryke, P.A.J. (1982). *Dierkunde: 'n Funksionele Benadering*. Durban: Butterworth Publishers (Pty) Ltd.

Sparks, J. (1982). *The Discovery of Animal Behaviour*. London: William Collins Sons & Co Ltd.

Vermes, G. (1977). *An Introduction to the Complete Dead Sea Scrolls*. London: SCM Press.

Wegener, G.S. (1958). *In die Begin was die Woord*. Cape Town: Human & Rousseau.

Wurthwein, E. (1995). *The Text of the Old Testament*. Grand Rapids: William B. Eerdmans Publishing Company.

Egyptian Vulture *(Neophron percnopterus)*
(Photo: Arturo) [4]

---

[4] This image is a reproduction of the photograph taken by Arturo, from Galicia (NW coast of Spain) and used with permission. Originally placed on www.flickr.com, the original photograph can be viewed at: http://www.flickr.com/photos/chausinho/4601495497/in/set-72157600950852011.

BIRDS AND BIBLES
IN HISTORY

PART 5
APPENDIXES

---

*Including 409 biblical references to birds
and avian terminology per Bible book and
per species for the benefit of scholars and laymen alike.*

Great Blue Heron (*Ardea herodias*)
(*Photo: David Dillon*) [1]

---

[1] This image is a reproduction of the photograph taken by David Dillon. Originally placed on www.flickr.com, the original photograph can be viewed at: http://www.flickr.com/photos/crotach/3849634767/sizes/l/in/set-72157612944891118/

CHAPTER 30
CLASSIFICATION:
THE LIVING WORLD

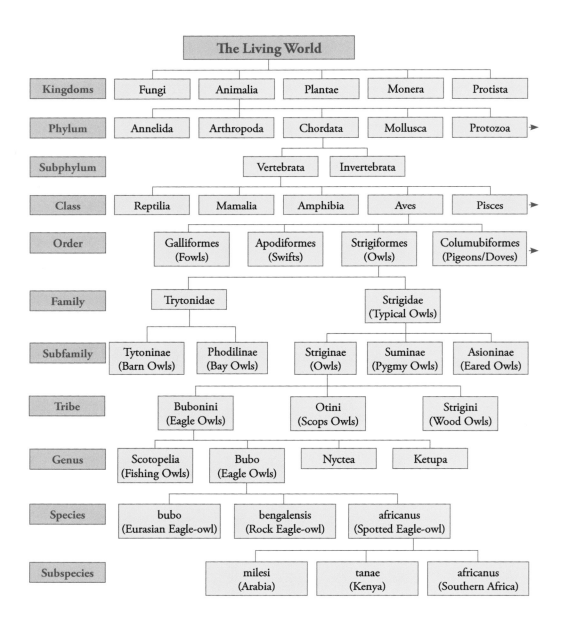

CHAPTER 31
CLASSIFICATION:
BUBO AFRICANUS

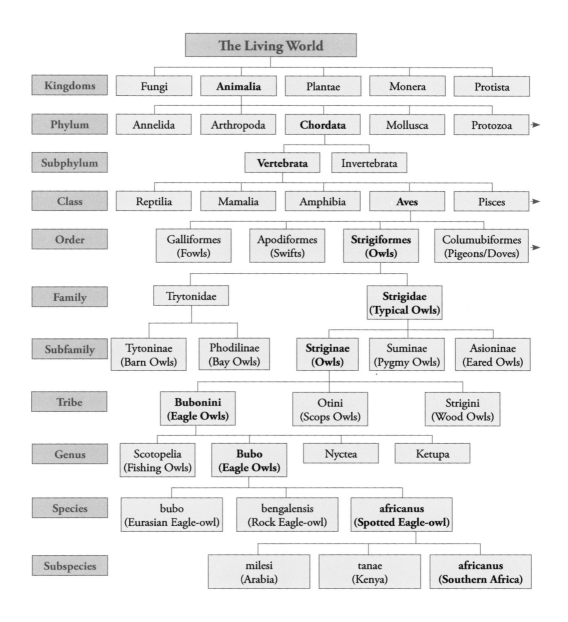

CHAPTER 32
REFERENCES TO BIRDS
PER BIBLE BOOK

NOTES:

1. In many modern translations the division of chapters into verses on occasion differs from the Masoretic Hebrew text. If it does not readily correspond, the reader has to look for the verse a few numbers up or down in the Bible that s/he is using.
2. The references are arranged from top to bottom in the columns, and then left to right.
3. The following references can all be found in this Chapter 32:

   ▪ Old Testament: Included are 282 references to birds. These include 174 references were a decision to a specific bird can be made, and 108 references to "birds" and "birds of prey" where it is impossible to be any clearer.
   ▪ New Testament: Included are 47 direct references to birds, but it is not possible to identify any specific species.

4. A summary of the other chapters:

	OT	NT	Total	Chapters
Identified species	174	0	174	35 36 (Also listed in this Chapter 32)
Refer to birds in general	108	47	155	34 (Also listed in this Chapter 32)
Avian terms	61	19	80	33
Total	**343**	**66**	**409**	

32.1 Old Testament books

Genesis

1:20	Birds	7:8	Birds	8:17	Birds
1:21	Birds	7:14	Birds	8:19	Birds
1:22	Birds	7:14	Birds	8:20	Birds
1:26	Birds	7:21	Birds	9:2	Birds
1:28	Birds	7:23	Birds	9:10	Birds
1:30	Birds	8:7	Raven	15:9	Turtle Dove
2:19	Birds	8:8	Rock Dove	15:10	Birds
2:20	Birds	8:9	Rock Dove	15:11	Birds of prey
6:7	Birds	8:10	Rock Dove	40:17	Birds
6:20	Birds	8:11	Rock Dove	40:19	Birds
7:3	Birds	8:12	Rock Dove		

Exodus

16:13	Quail
19:4	Eagle

Leviticus

1:14	Birds	11:17	Little Owl (kos)	14:6	Birds
1:14	Turtle Dove	11:17	Cormorant	14:7	Birds
5:7	Rock Dove	11:17	Long-eared Owl	14:22	Rock Dove
5:7	Turtle Dove	11:18	Barn Owl	14:22	Turtle Dove
5:11	Rock Dove	11:18	Little Owl (qa'ath)	14:30	Rock Dove
5:11	Turtle Dove	11:18	Egyptian Vulture	14:30	Turtle Dove
7:26	Birds	11:19	Stork	14:49	Birds
11:13	Birds	11:19	Heron,	14:50	Birds
11:13	Eagle	11:19	Hoopoe	14:51	Birds
11:13	Bearded Vulture	11:19	Bat	14:52	Birds
11:13	Lapped-faced Vulture	11:20	Bird	14:53	Birds
11:14	Red Kite	11:46	Birds	15:14	Rock Dove
11:14	Black Kite	12:6	Rock Dove	15:14	Turtle Dove
11:15	Raven	12:6	Turtle Dove	15:29	Rock Dove
11:16	Ostrich (daughter of)	12:8	Rock Dove	15:29	Turtle Dove
11:16	Scops Owl	12:8	Turtle Dove	17:7	Scops Owl
11:16	Sea-gull	14:4	Birds	17:13	Birds
11:16	Falcon	14:5	Birds	20:25	Birds

Numbers

6:10	Rock Dove	11:31	Quail
6:10	Turtle Dove	11:32	Quail

Deuteronomy

4:17	Birds	14:15	Ostrich (daughter of)	14:17	Cormorant
14:11	Birds	14:15	Scops Owl	14:18	Stork
14:12	Eagle	14:15	Sea-gull	14:18	Heron
14:12	Bearded Vulture	14:15	Falcon	14:18	Hoopoe
14:12	Lapped-faced Vulture	14:16	Little Owl *(kos)*	14:18	Bat
14:13	Red Kite	14:16	Long-eared Owl	14:20	Birds
14:13	Black Kite	14:16	Barn Owl	28:26	Birds
14:13	Griffon Vulture	14:17	Little Owl *(qa'ath)*	28:49	Eagle
14:14	Raven	14:17	Egyptian Vulture	32:11	Eagle

1 Samuel

17:44	Birds
17:46	Birds
26:20	Partridge

2 Samuel

1:23	Eagle

1 Kings

4:33	Birds
5:3	Lark-heeled Cuckoo
10:22	Indian Peafowl
17:4, 6	Raven

2 Kings

23:8	Scops Owl

2 Chronicles

9:21	Indian Peafowl
11:15	Scops Owl

Job

9:26	Eagle	35:11	Birds	39:13	Stork
15:23	Black Kite ?	38:36	Cock	39:26	Falcon
28:7	Birds of prey	38:36	Ibis	39:26	Bird
28:7	Black Kite	38:41	Raven	39: 27	Eagle
28:21	Birds	39:13	Ostrich, female	40:29	Birds/Plover
30:29	Ostrich				

Psalms

11:1	Bird	79:2	Birds	104:12	Birds
50:11	Birds	84:4	Bird	104:17	Stork
55:7	Rock Dove	84:4	Swallow	104:17	Bird
56:1	Rock Dove	102:7	Little Owl	105:40	Quail
68:14	Rock Dove	102:7	Little Owl	124:7	Bird
74:19	Turtle Dove	102:8	Bird	147:9	Raven
78:27	Birds	103:5	Eagle		

Proverbs

1:17	Birds	26:2	Swallow	30:17	Raven
6:5	Bird	26:2	Bird	30:17	Vulture
7:23	Bird	27:8	Bird	30:19	Eagle
23:5	Eagle				

Ecclesiastes

5:11	Raven	9:12	Bird	10:20	Birds
12:4	Bird				

Song of Songs

1:13	Rock Dove	2:14	Rock Dove	5:11	Raven
1:15	Rock Dove	4:1	Rock Dove	5:12	Rock Dove
2:12	Turtle Dove	5:2	Rock Dove	6:9	Rock Dove

Isaiah

2:20	Bat	34:11	Little Owl (qa'ath)	38:14	Swift
13:21	Eagle Owl	34:11	Short-eared Owl	38:14	Bulbul
13:21	Ostrich	34:11	Long-eared Owl	38:14	Rock Dove
13:21	Scops-Owl	34:11	Raven	40:31	Eagle
14:23	Short-eared Owl	34:13	Ostrich	43:20	Ostrich
16:2	Birds	34:14	Scops Owl (sa'ir)	46:11	Birds of prey
18:6	Birds of prey	34:15	Tawny Owl	59:11	Rock Dove
18:6	Birds of prey	34:15	Griffon Vulture	60:8	Rock Dove
31:5	Bird				

Jeremiah

4:13	Eagle	9:10	Birds	19:7	Birds
4:25	Birds	12:4	Birds	34:20	Birds

5:27	Birds	12:9	Spotted Birds of Prey	48:28	Rock Dove
7:33	Birds	12:9	Birds of Prey	48:40	Eagle
8:7	Stork	15:3	Birds	49:16	Eagle
8:7	Turtle Dove	16:4	Birds	49:22	Eagle
8:7	Swift	17:11	Partridge	50:39	Ostrich
8:7	Bulbul				

Lamentations

| 3:52 | Birds | 4:3 | Ostrich | 4:19 | Eagle |

Ezekiel

1:10	Eagle	17:23	Bird	32:4	Birds
7:16	Rock Dove	29:5	Birds	38:20	Birds
10:14	Eagle	31:6	Birds	39:4	Birds of prey
17:3	Eagle	31:13	Birds	39:17	Bird
17:7	Eagle				

Daniel

2:38	Birds	4:21	Bird	7:4	Eagle
4:12	Bird	4:30	Eagle	7:6	Birds
4:14	Bird	4:30	Bird		

Hosea

2:17	Birds	7:12	Birds	11:11	Rock Dove
4:3	Birds	9:11	Birds	11:11	Ostrich
7:11	Rock Dove	11:11	Birds		

Amos

| 3:5 | Birds |

Obadiah

| 1:4 | Eagle |

Micah

| 1:8 | Ostrich |
| 1:16 | Eagle |

Nahum

| 2:7 | Dove |

Habakkuk

| 1:8 | Eagle |

Zechariah

| 5:9 | Stork |

Zephaniah

| 1:3 | Birds | 2:14 | Little Owl (qa'ath) | 2:14 | Short-eared Owl |

32.2 New Testament books

Matthew

3:16	Dove	10:31	Sparrow	24:28	Vultures
6:26	Birds	13:4	Birds	26:34	Cock
8:20	Birds	13:32	Birds	26:75	Cock
10:29	Sparrow	21:12	Dove	26:75	Cock
10:16	Dove	23:37	Hen		

Mark

1:10	Dove	11:15	Dove	14:68	Cock
4:4	Birds	13:35	Crowing	14:72	Cock
4:32	Birds	14:30	Cock		

Luke

2:24	Turtle Dove	12:7	Sparrow	22:34	Cock
2:24	Dove	12:24	Raven	22:60	Cock
3:22	Dove	13:34	Hen	22:61	Cock
12:6	Sparrow	17:37	Vultures		

John

1:32	Dove	2:16	Dove	18:27	Cock
2:14	Dove	13:38	Cock		

Acts of the Apostles

10:12	Birds
11:6	Birds

Romans

1:23	Birds

1 Corinthians

15:39	Birds

Revelation

4:7	Eagle	8:13	Eagle	12:14	Eagle
19:17	Birds of prey	19:21	Birds of prey		

CHAPTER 33
REFERENCES TO AVIAN TERMS

33.1 Old Testament

1. Chicks

Gen 15:9	Deut 32:11	Prv 30:17

2. Nest

Deut 22:6	Deut 32:11	Jer 49:16

3. Feather

Lev 1:16	Job 39:13	Job 39:18	Ps 68:13	Ps 78:27	Ps 91:4
Ezk 10:12	Ezk 17:3	Ezk 17:7	Ezk 39:4	Ezk 39:17	Dan 4:33

4. Fly

Prv7:23	Prv 23:5	Prv26:2	Is 31:5	Hos 9:11

5. Kind

Gen 1:11	Gen 6:20	Gen 7:14	Gen 21:24	Lev 11:14	Lev 11:15
Lev 11:16	Lev 11:19	Lev 11:22	Lev 11:29	Deut 14:13	Deut 14:14
Deut 14:15	Deut 15:18	Lam 47:10			

6. Wing

Ex 19:4	Lev 1:17	Job 38:26	Dan 7:4	Dan 7:6	Is 40:31 Ezk
10:12					

7. Bird catcher

Ps 124:7	Prv 6:5	Jer 5:26.	Jer 5:27

8. Egg

Deut 22:6	Job 6:6	Job 39:14	Is 10:14	Is 34:15	Is 59:5
Jer 17:11					

9. Scarecrow	10. Soar	11. Crop	12. Fly away	13. Unexplained
Jer 10:5	Job 38:26	Lev 1:16	Jer 4:25	Ezk 13:20

33.2 New Testament

1. Birds

Mat 6:26	Mat 8:20	Mat 13:4	Mat 13:32	Mrk 4:4	Mrk 4:32
Luk8:5	Luk 9:58	Luk 12:24	Luk 13:19	Act 10:12	Act 11:6
Rom 1:23	I Cor 15:39				

2. Wing	3. Egg	4. Crowing	5. Birds of prey	
Rev 12:14	Luk 11:12	Mrk 13:35	Rev 19:17	Rev 19:21

Canada Goose (*Branta canadensis*)
(*Photo: Tony Hisgett*) [1]

---

[1] This image is a reproduction of the photograph taken by Tony Hisgett, originally from Huddersfield, Yorkshire, England and used with permission. Originally placed on www.flickr.com, the original photograph can be viewed at: http://www.flickr.com/photos/hisgett/3543279918/.

CHAPTER 34
REFERENCES TO BIRDS IN GENERAL

34.1 Old Testament

1. Birds (*'oph*)

Gen 1:20	Gen 7:21	Lev 11:20	Deut 4:17	Ps 104:12	Ezk 31:6
Gen 1:21	Gen 7:23	Lev 11:46	Deut 14:11	Ecl 10:20	Ezk 31:13
Gen 1:22	Gen 8:17	Lev 14:4	Deut 14:20	Is 16:2	Ezk 32:4
Gen 1:26	Gen 8:19	Lev 14:5	Deut 28:26	Jer 4:25	Ezk 38:20
Gen 1:28	Gen 8:20	Lev 14:6	1 Sam 17:44	Jer 5:27	Dan 2:38
Gen 1:30	Gen 9:2	Lev 14:7	1 Sam 17:46	Jer 7:33	Dan 7:4
Gen 2:19	Gen 9:10	Lev 14:49	1 Kng 4:33	Jer 9:10	Hos 2:17
Gen 2:20	Gen 15:10	Lev 14:50	Job 28:21	Jer 12:4	Hos 4:3
Gen 6:7	Gen 40:19	Lev 14:51	Job 35:11	Jer 15:3	Hos 7:12
Gen 6:20	Gen 40:17	Lev 14:52	Job 40:29	Jer 16:4	Hos 9:11
Gen 7:3	Lev 1:14	Lev 14:53	Ps 50:11	Jer 19:7	Zeph 1:3
Gen 7:8	Lev 7:26	Lev 17:13	Ps 78:27	Jer 34:20	
Gen 7:14	Lev 11:13	Lev 20:25	Ps 79:2	Ezk 29:5	

2. Birds; Winged animals (*tsippor*)

Gen 7:14	Ps 104:17	Prv 26:2	Is 31:5	Dan 4:12	Hos 11:11
Ps 11:1	Ps 124:7	Prv 27:8	Lam 3:52	Dan 4:14	Am 3:5
Ps 84:4	Prv 6:5	Ecl 9:12	Ezk 17:23	Dan 4:21	
Ps 102:8	Prv 7:23	Ecl 12:4	Ezk 39:17	Dan 4:30	

NOTE: The name of Moses' wife is derived from the same stem *tsippor*. Zipporah or Tsipporah then means "female bird."

3. Birds (*kanaph*)

Job 39:26	Prv 1:17

4. Birds of prey *('ajit)*

Gen 15:11	Is 18:6	1s 46:11	Jer 12:9
Job 28:7	Is 18:6	Jer 12:9	Ezk 39:4

Southern Red Bishop *(Euplectes orix)*
(Photo: Rus Koorts) [1]

---

[1] This image is a reproduction of the photograph taken by Rus Koorts from Pretoria, in South Africa and used with permission. Originally placed on www.flickr.com, the original photograph can be viewed at: http://www.flickr.com/photos/ruslou/6301262854/in/set-72157625798426776

CHAPTER 35
BIBLE SPECIES ALPHABETICAL

NOTE: In order to keep them alphabetically in the broad vernacular groups, parts of some names have been separated and re-arranged. For example: "Pallid Scops-owl" changed to "Owl, Scops, Pallid."

35.1 Old Testament

01.	*'atalleph*	Bats	(Chiroptera)
02.	*'agur*	Bulbul, Yellow-vented	(Pycnonotus xanthopygos)
03.	*Barbur*	Cuckoos	(Cuculidae)
04.	*Shalah*	Cormorant	(Phalacrocorax carbo)
05.	*Jonah*	Dove, Rock	(Columbia livia)
06.	*Tor*	Dove, Turtle	(Streptopelia turtur)
07.	*Nesher*	Eagles	(Accipitridae)
08.	*Nets*	Falcon, Peregrine	(Falco peregrines)
09.	*'anaphah*	Herons	(Ardeidae)
10.	*Dukhiphath*	Hoopoe	(Upupa epops)
11.	*Sekwi*	Junglefowl, Red	(Gallus gallus)
12.	*'ajjah*	Kite, Black	(Milvus migrans migrans)
13.	*Da'ah*	Kite, Red	(Milvus milvus milvus)
14.1	*Ya'anah*	Ostrich	(Struthio camelus)
14.2	*Renanim*	Ostrich, female	(Struthio camelus)
15.	*Tinshemet*	Owl, Barn, common	(Tyto alba)
16.	*Och*	Owl, Eagle	(Bubo bubo)
17.	*Kos*	Owl, Little	(Athene noctua saharae)
18.	*Qa'ath*	Owl, Little	(Athene noctua lilith)
19.	*Yanshoph*	Owl, Long-eared, N	(Asio otus)
20.	*Sa'ir*	Owl, Scops, Eurasian	(Otus scops scops)
21.	*Tahmas*	Owl, Scops, Pallid	(Otus brucei)
22.	*Qippod*	Owl, Short-eared	(Asio flammens)
23.	*Qippoz*	Owl, Tawny	(Strix aluco)

24.	*Qore'*	Partridge	*(Ammoperdrix hayi)*
25.	*Tukkijjim*	Peafowl, Indian	*(Pavo cristatus)*
26.	*Stippor*	Plover, Egyptian	*(Pluvianus aegyptius)*
27.	*Slaw*	Quail	*(Cotumix cotumix)*
28.	*'orev*	Ravens	*(Corvidae)*
29.	*Shahaph*	Sea-gulls	*(Laridae)*
30.	*Hasidhah*	Storks	*(Cicinidae)*
31.	*Dror*	Swallows	*(Hirundinidae)*
32.	*Sis*	Swift, Common	*(Apus apus)*
33.	*Peres*	Vulture, Bearded	*(Gypaetus barbatus)*
34.	*Dajjah*	Vulture, Griffon	*(Gyps fulvus)*
35.	*Raham*	Vulture, Egyptian	*(Neophron percnopterus)*
36.	*'oznijah*	Vulture, Lappet-faced	*(Torgos tracheliotus negevensis)*

35.2 New Testament

alektor	Cock
peristera	Dove
aetos	Eagle
ornis	Hen, domestic
koraks	Raven
strution	Sparrow
trugon	Turtle Dove
aetos	Vulture

Southern Grey Shrike *(Lanius meridionalis)*
(Photo: Dr. Marcel Holyoak) [1]

---

[1] This image is a reproduction of the photograph taken by Dr. Marcel Holyoak and used with permission. Originally placed on www.flickr.com, the original photograph can be viewed at: http://www.flickr.com/photos/maholyoak/5918071164/sizes/l/in/photostream/.

CHAPTER 36
REFERENCES AS PER BIBLE SPECIES

NOTES:

1. In many modem translations the division of chapters into verses on occasion differs from the Masoretic Hebrew text.
2. In order to keep the names alphabetically in their broad vernacular groups, the parts of some names have been separated and re-arranged arbitrarily. For example: "Pallid Scops-owl" has been changed to "Owl, Scops, Pallid."
3. The references are listed from left to right in the columns.

36.1 Old Testament

1. Bat (Hebrew: *'atalleph*) *(Chiroptera Spp.)*
Lev 11:19   Deut l4:18   Is 2:20

2. Bulbul, Yellow-vented (Hebrew: *'agur*) *(Pycnonotus xanthopygos)*
Is 38:14   Jer 8:7

3. Cuckoo, Lark-heeled (Hebrew: *barbur*) *(Centropus aegyptius)*
1 Kng 5:3

4. Cormorant (Hebrew: *shalah*) *(Phalacrocorax carbo)*
Lev 11:17   Deut l4:17

5. Dove, Rock (Hebrew: *jonah*) *(Columbia livia)*

Gen 8:8	Gen 8:9	Gen 8:10	Gen 8:11	Gen 8:12	Lev5:7
Lev 5:11	Lev 12:6	Lev l2:8	Lev 14:22	Lev 14:30	Lev 15:14
Lev 15:29	Num 6:10	Ps 55:7	Ps 56:1	Ps 68:14	So 1:13
So 1:15	So 2:14	So 4:1	So 5:2	So 5:12	So 6:9

Is 38:14   Is 59:11   Is 60:8   Jer 48:28   Hos7:11   Ho 11:11
Na 2:7   Ezk 7:16

6. Dove, Turtle (Hebrew: *tor*) *(Streptopelia turtur)*
Gen 15:9   Lev 1:14   Lev 5:7   Lev 5:11   Lev 12:6   Lev 12:8
Lev 14:22   Lev 14:30   Lev 15:14   Lev 15:29   Num 6:10   Ps 74:19
So 2:12   Jer 8:7

7. Eagle (Hebrew: *nesher*) *(Accipitridae Spp.)*
Ex19:4   Lev 11:13   Deut 14:12.   Deut 28:49   Deut 32:11   2 Sam 1:23
Job 9:26   Job 39:27   Ps 103:5   Prv 23:5   Prv 30:19   Is 40:31 Jer
4:13   Jer 48:40   Jer 49:16   Jer 49:22   Lam 4:19   Ezk 1:10
Ezk 10:14   Ezk 17:3   Ezk 17:7   Dan 4:30   Dan 7:4   Obd 1:4
Mi 1:16   Hab 1:8

8. Falcon (Hebrew: *nets*) *(Falco peregrinus)*
Lev 11:16   Deut l4:15   Job 39:26

9. Heron (Hebrew: '*anaphah*) *(Ardeidae Spp.)*
Lev 11:19   Deut l4:18

10. Hoopoe (Hebrew: *dukhiphath*) *(Upupa epops)*
Lev 11:19   Deut 14:18

11. Junglefowl, Red (Hebrew: *sekwi*) *(Gallus gallus)*
Job 38:36

12. Kite, Black (Hebrew: '*ajjah*) *(MiIvus migrans migrans)*
Lev 11:14   Deut l4:13   Job 15:23 ?   Job 28:7

13. Kite, Red (Hebrew: *da'ah*) *(Milvus milvus milvus)*
Lev 11:14   Deut 14:13

14.1 Ostrich (Hebrew: *ja'anah*) *(Struthio camelus)*
Lev 11:16   Deut l4:15   Job 30:29   Is 13:21   Is 34:13   Is 43:20
Jer 50:39   Lam 4:3   Hos 11:11   Mi 1:8

14.2. Ostrich, female (Hebrew: *renanim*) *(Struthio camelus)*
Job 39:13

15. Owl, Common Barn (Hebrew: *tinshemet*) *(Tyto alba)*
Lev 11:18   Deut 14:16

16. Owl, Eagle (Hebrew: *och*) *(Bubo bubo)*
Is 13:21

17. Owl, Little (Hebrew: *kos*) *(Athene noctua saharae)*
Lev 11:17   Deut 14:16   Ps 102:7

18. Owl, Little (Hebrew: *qa'ath*) *(Athene noctua lilith)*
Lev 11:18   Deut 14:17   Ps 102:7   Is 34:11   Zeph 2:14

19. Owl, Long-eared, Northern (Hebrew: *janshoph*) *(Asio otus)*
Lev 11:17   Deut 14:16   Is 34:11

20. Owl, Scops, Eurasian (Hebrew: *sair*) *(Otus scops scops)*
Lev l7:7   2 Kng 23:8   2 Chr 11:15   Is 13:21   Is 34:14

21. Owl, Scops Pallid (Hebrew: *tahmas*) *(Otus brucei)*
Lev 11:16   Deut 14:15

22. Owl, Short-eared (Hebrew: *qippod*) *(Asio flammeus)*
Is 14:23   Is 34:11   Zeph 2:14

23. Owl, Tawny (Hebrew: *qippoz*) *(Strix aluco)*
Is 34:15

24. Partridge (Hebrew: *qore*) *(Ammoperdrix hayi)*
1 Sam 26:20   Jer 17:11

25. Peafowl, Indian (Hebrew: *tukkijjim*) *(Pavo cristatus)*
1 Kng 10:22   2 Chr 9:21

26. Plover, Egyptian (Hebrew: *stippor*) *(Pluvianus aegyptius)*
Job 40:29

27. Quail (Hebrew: *slaw*) *(Coturnix coturnix)*
Ex 16:13   Num 11:31   Num 11:32   Ps 105:40

28. Raven (Hebrew: '*orev*) *(Corvus Spp.)*
Gen 8:7   Lev 11:15   Deut 14:14   1 Kng 17:4   I Kng 17:6   Job 38:41
Ps 147:9   Prv 30:17   Ecl 5:11   So 5:11   Is 34:11

29. Sea-gull (Hebrew: *shahaph*) *(Laridae)*
Lev 11:16   Deut 14:15

30. Stork (Hebrew: *hasidhah*) *(Ciconidae)*
Lev 11:19   Deut 14:18   Job 39:13   Ps 104:17   Jer8:7   Zech 5:9

31. Swallow (Hebrew: *dror*) *(Hirundinidae Spp.)*
Ps 84:4   Prv 26:2

32. Swift (Hebrew: *sis*) *(Apus apus)*
Is 38:14   Jer 8:7

33. Vulture, Bearded (Hebrew: *peres*) *(Gypaetus barbatus)*
Lev 11:13   Deut 14:12

34. Vulture, Griffon (Hebrew: *dajjah, nesher*) *(Gyps fulvus)*
Deut l4:13   Is 34:15   Mi 1:16   Prv 30:17

35. Vulture, Lapped-faced (Hebrew: *oznijah*) *(Torgos tracheliotus negevensis)*
Lev 11:13   Deut l4:12

36. Vulture, Egyptian (Hebrew: *raham*) *(Neophron percnopterus)*
Lev 11:18   Deut l4:17

36.2 New Testament

1. Dove (Greek: *peristera*)
Mat 3:16   Mat 10:16   Mat 21:12   Mrk 1:10   Mrk 11:15   Luk 2:24
Luk 3:22   Jhn 1:32   Jhn 2:14   Jhn 2:16

2. Cock (Greek: *alektor*)

Mat 26:34	Mat 26:75	Mat 26:75	Mrk l4:30	Mrk14:68	Mrk 14:72
Luk 22:34	Luk 22:60	Luk 22:61	Jhn 13:38	Jhn 18:27	

3. Eagle (Greek: *aetos*)

Rev 4:7   Rev 8:13   Rev 12:14

4. Hen, domestic (Greek: *ornis*)

Mat 23:37   Luk 13:34

5. Turtle Dove (Greek: *trugon*)

Luk2:24

6. Raven (Greek: *koraks*)

Luk 12:24

7. Sparrow (Greek: *strution* **which means "assortment of small birds")**

Mat 10:29   Mat 10:31   Luk 12:6   Luk 12:7

8. Vulture (Greek: *aetos*)

Mat 24:28   Luk 17:37

Mourning Wheatear *(Oenanthe lugens)*
(Photo: Dr. Marcel Holyoak) [1]

---

[1] This image is a reproduction of the photograph taken by Dr. Marcel Holyoak and used with permission. Originally placed on www.flickr.com, the original photograph can be viewed at: http://www.flickr.com/photos/maholyoak/5918023712/.

CHAPTER 37
INDEX TO SPECIES AND FAMILIES

NOTES TO THIS INDEX:

1. In order to keep the names alphabetically in their broad vernacular groups, the parts of some names have been separated and re-arranged arbitrarily. For example: "Pallid Scops-owl" has been changed to "Owl, Scops, Pallid", and "African Little Sparrow-hawk" to "Hawk, Sparrow, African Little."
2. In order to have a single reference per name, those few that are mentioned more than once have been duplicated.

37.1 Old Testament species and families

No.	English Name	References to species in this publication						
	Name	Par.	Par.	Par.	Par.	Par.	Par.	Par.
01.	Bats	2.4	2.6	10.1	17.1	17.4	26.1	26.1.1
02.	Bulbul, Y-vented	4.7	26.2.3	35.1				
03.	Cuckoos	1.1	1.6.2	2.5	3.5	4.7	9.3	26.2.1
04.	Cormorant	3.4	3.6	4.7	26.2.2			
05.	Dove, Rock	18.1	19.1	19.2	19.3	35.1		
06.	Dove, Turtle	19.3	35.1	35.2				
07.	Eagles	3.1	4.2.3	4.8	9.1	9.3	10.1	10.1.1
		11.2	12.1	26.2.3				
08.	Falcon, Peregrine	11.1	11.3	35.1				
09.	Herons	3.1	4.7	26.2.2	35.1			
10.	Hoopoe	4.7	26.2.2	35.1				
11.	Junglefowl, Red	3.5	14.4	14.4.1	35.1			
12.	Kite, Black	9.3	10.3	10.3.1	35.1			

13.	Kite, Red	10.3	10.3.1	26.2.2	35.1			
14.1	Ostrich	1.3	3.5	4.7	7.1	8.1	8.2	8.3
		10.2	27.3	35.1				
14.2	Ostrich, female	8.1	8.3	35.1				
15.	Owl, Barn, common	1.5	2.1.3	2.4	4.7	15.1	15.2	16.1
		16.1	16.2	16.3	26.2.2	35.1		
16.	Owl, Eagle	1.6.1	4.6	4.6.1	4.6.2	4.7	17.1	17.2
		26.2.1	26.2.3	30	31	35.1		
17.	Owl, Little (Sahara)	15.1	17.5	17.5.1	26.3.1	35.1		
18.	Owl, Little (Lilith)	15.1	17.5	17.5.1	26.3.1	35.1		
19.	Owl, Long-eared, N	17.6	17.6.1	17.7	17.7.1	26.3.1	35.1	
20.	Owl, Scops, Europe	15.1	17.3	26.2.1	35.1			
21.	Owl, Scops, Pallid	17.4	26.2.2	35.1				
22.	Owl, Short-eared	17.7	17.7.1	35.1				
23.	Owl, Tawny	17.8	26.2.1	35.1				
24.	Partridges	1.3	13.1	14.1	14.2	14.2.1	26.2.3	35.1
25.	Peafowl, Indian	1.3	1.6.1	3.5	14.1	14.5	14.5.1	
		26.2.3	35.1					
26.	Plover, Egyptian	26.3.2	35.1					
27.	Quail	4.7	13.1	13.2	14.1	14.3	14.3.1	35.1
28.	Ravens	17.6	19.3	20.1	22.1	22.2	22.3	
		26.3.1	27.1	27.2	35.1			
29.	Sea-gulls	2.2	3.5	4.7	26.2.2	35.1	40.6	
30.	Storks	1.6.1	4.2	4.4	4.7	4.8	26.3.3	35.1
31.	Swallows	1.3	2.2	2.4	4.2	4.7	21.1	21.2
		26.2.3	35.1					

32.	Swift, Common	2.4	4.2	4.4	4.5	4.7	11.1	21.1
		24.1	25.1	25.2	26.2.3			
33.	Vulture, Bearded	10.2	10.2.1	26.2.2	35.1			
34.	Vulture, Griffon	9.3	10.1	10.2	10.2.1	26.2.1	26.2.3	35.1
35.	Vulture, Egyptian	6.3	9.3	10.2	10.2.1	12.3	26.2.2	35.1
36.	Vulture, Lappet-faced	10.2	10.2.1	26.2.2	35.1			

Great Blue Heron *(Ardea herodias)*
(Photo: David Hofmann) [1]

37.2 Other Species and Families

Name	Par.	Name	Par.
Auk, Great	3.1	Buzzard	9.3
Avocet	3.5	Buzzard, Honey	4.8
Babbler	3.5, 6	Canaries	3.6
Babbler, Arabian	3.6	Caracaras	9.3
Barbets	17.1	Cassowaries	7.1
Bazas	9.3	Chachalacas	13.1, 2
Birds of Paradise	20.1	Chaffinches	3.4
Bittern	26.3.1	Chickens	3.6
Bowerbirds	20.1	Chickens	13.1
Bullfinches	2.5	Chiffchaff	2.4
Bustard	3.5	Choughs	22.1
Bustard, Great	26.3.1	Chukar	14.1, 2
Buzzard	4.8	Cock	13.1
		Cock	14.4
		Condor	9.3
		Condor	10.2
		Condor, Andean	9.2
		Condor, California	9.2
		Condor, California	10.2
		Corvids	17.2
		Crakes	1.4
		Crane	3.5
		Crow	1.3
		Crow	4.6.3
		Crow	17.6
		Crow	22.1, 3
		Crow, Hooded	22.2
		Crow, Carrion	22.2
		Crow, Bush, Stresemann's	22.1
		Cranes	26.3.3
		Cuckoo	1.1

Common Buzzard *(Buteo buteo)*
(Photo: Ted van den Bergh) [2]

---

[2] This image is a reproduction of the photograph taken by Ted van den Bergh and used with permission. Originally placed on www.flickr.com, the original photograph can be viewed at: http://www.flickr.com/photos/webted/4629819539/.

Name	Par.	Name	Par.
Cuckoo	3.5	Eagle, Fish	10.1
Cuckoo, Common	1.6.2	Eagle, Golden	9.3
Cuckoo, Lark-heeled	26.2.1	Eagle, Golden	10.1
Curassows	13.1, 2	Eagle, Hawk	9.3
Curlew, Eskimo	3.1	Eagle, Short-toed	9.3
Curlew, Esquimaux	3.1	Eagle, Spotted	9.3
Curlew, Northern	3.1	Eagle, Spotted	10.1
Dippers	20.1	Eagle, Steppe	10.1
Dodo	2.1.3, 4	Eagle, Sea	9.3
Dodo	4.5	Eagle, Snake	9.3
Dodo	18.1	Eagle, Snake, Old World	10.1
Dove	1.1	Eagle, Serpent	9.3
Dove	4.4	Eagle, Serpent, New World	10.1
Dove	17.1	Eagle, True	9.3
Dove	18.1	Eider	3.5
Dove	19.1	Elephant bird	7.1
Dove, Mourning, American	4.5	Emu	7.1
Dove, Collared	19.3	Falcon	9.1, 3
Dove, Eared	4.5	Falcon	11.1
Dove, Fruit	18.1	Falcon, Barbary	11.1
Dove, Ground, Bare-faced	19.1	Falcon, Eleonora's	11.1
Dove, Ground, Common	18.1	Falcon, Forest	9.3
Dove, Laughing	19.3	Falcon, Forest	11.1
Dove, Namaqua	19.1, 3	Falcon, Lanner	11.1
Dove, Palm	19.3	Falcon, Lanner	11.2
Duck	4.2	Falcon, Peregrine	11.1
Duck	17.2	Falcon, Peregrine	11.3
Duck, Labrador	3.1	Falcon, Pygmy, African	11.1
Duck, Mandarin	1.4	Falcon, Saker	11.2
Eagle	4.8	Falcon, Taita	4.1
Eagle	9.1	Falcon, Taita	11.1
Eagle	10.1	Falconets	9.3
Eagle	11.2	Falconets	11.1
Eagle, Bald	10.1	Finches	3.6
Eagle, Booted	9.3	Fowl, mallee	13.1
Eagle, Bonnelli's	10.1	Francolin	1.3
Eagle, Fish	9.3	Francolin	14.1

Name	Par.	Name	Par.
Francolin, Black	14.1	Heron, Night	26.3.1
Gannets	3.6	Heron, Purple	26.2.2
Geese	17.2	Hummingbirds	24.1
Geese, Barnacle	1.6.2	Icterids	20.1
Geese, Greylag	3.4, 5	Ibis	14.4.1
Goldeneyes	3.5	Ibis	26.2.1
Goshawk	17.2	Jackdaw	3.5
Goshawk	9.3	Jackdaw	22.2
Goshawk	10.1	Jay, Grey	22.1
Goshawk, Chanting	9.3	Jay, New World	22.1
Grebe, Crested, Great	3.5	Jay, Old World	22.1
Grouse	13.1, 2	Kestrel	9.3
Grouse	17.2	Kestrel	11.1
Grouse, Pinnated	3.1	Kestrel, Australian	11.1
Guans	13.1, 2	Kestrel, Common	9.3
Guineafowl	13.1, 2	Kestrel, Common	11.1
Gull, Herring	3.5	Kestrel, Lesser	11.1
Gymnogene	10.1	Kestrel, Mauritius	11.1
Hamerkop	17.1	Kingfishers	12.1
Harrier	9.3	Kite	9.3
Harrier, Marsh	9.3	Kite	10.3
Hawk	9.1, 3	Kite, Black	9.3
Hawk	11.2	Kite, Black	10.3
Hawk	17.6	Kite, Black	10.3.1
Hawk, Bat	10.1	Kite, Black-shouldered	10.3
Hawk, Cuckoo	9.3	Kite, Black-winged	10.3
Hawk, Fish	12.1	Kite, Red	10.3
Hawk, Harrier	9.3	Kite, Red	10.3.1
Hawk, Harrier, African	10.1	Kittiwakes	3.5
Hawk, Sparrow	9.3	Kiwi	7.1
Hawk, Sparrow	10.1	Koel, Asian	1.1
Hawk, Sparrow, African Little	9.3	Lammergeier	10.2.1
Hen	1.3	Lammergeyer	10.2.1
Hen	14.4.1	Lark, Crested	1.3
Heron, Blue, Great	3.1	Linnet	2.5
Heron, Goliath	26.2.2	Lyrebirds	20.1
Heron, Grey	26.2.2	Magpie	1.6.1

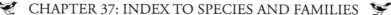

Name	Par.	Name	Par.
Magpie	17.6	Owl, Barnyard	16.1
Magpie, Azure-winged	22.1	Owl, Bay, Oriental	15.2
Magpie, Holarctic	22.1	Owl, Bay, Congo	15.2
Magpie, Oriental	22.1	Owl, Burrowing	15.1
Martin	2.4	Owl, Cave	16.1
Martin, Crag	21.2	Owl, Church	16.1
Martin, Crag, Pale	21.2	Owl, Death	16.1
Martin, House	21.2	Owl, Delicate	16.1
Martin, Rock	21.2	Owl, Demon	16.1
Martin, Sand	21.2	Owl, Dobby	16.1
Megapodes	13.2	Owl, Eagle	1.6.1
Mesites	13.1	Owl, Eagle	15.2
Moa	7.1	Owl, Eagle, Common	17.2
Mudnest Builders	20.1	Owl, Eagle, Eurasian	3.3
Mynahs	22.1		
Nightingale	2.2		
Nightjars	15.1		
Nightjars	25.1		
Nutcrackers	22.1		
Oriole, Black-naped	1.4		
Osprey	1.6.1		
Osprey	6.3		
Osprey	9.2, 3		
Osprey	12.1		
Osprey	12.2		
Osprey	12.3		
Ovenbirds	20.1		
Owl	1.1		
Owl	1.3		
Owl	2.7		
Owl	3.3		
Owl	15.1		
Owl, Barn, Typical	15.2		

Tawny Owl *(Asio flammeus)*
(Photo: Wiltshire Yan) [3]

---

[3] This image is a reproduction of the photograph taken by Wiltshire Yan and used with permission. Originally placed on www.flickr.com, the original photograph can be viewed at: http://www.flickr.com/photos/30945961@N05/4691282958/.

Name	Par.	Name	Par.
Owl, Eagle, Great	17.2	Owl, Scops, Seychelles	4.5
Owl, Eagle, Northern	17.2	Owl, Scops, Striated	17.4
Owl, Eagle, Pharaoh	17.2	Owl, Screech	1.5
Owl, Eagle, Spotted	4.6.1	Owl, Screech	15.2
Owl, Eagle, Spotted	4.6.2	Owl, Screech	16.1
Owl, Eagle, Spotted	4.6.3	Owl, Screech, Desert	17.4
Owl, Eagle, Spotted	17.1	Owl, Screech, Puerto Rican	15.1
Owl, Eared	15.2	Owl, Scritch	16.1
Owl, Elf	17.1	Owl, Short-eared, Hawaiian	17.7
Owl, Fish, Brown	17.8	Owl, Silver	16.1
Owl, Fish, Blakiston's	17.1	Owl, Snowy	15.1
Owl, Fishing	17.1	Owl, Sooty	15.2
Owl, Ghost	16.1	Owl, Stone	16.1
Owl, Golden	16.1	Owl, Straw	16.1
Owl, Golden, Masked	16.1	Owl, Tawny, Hume's	17.8
Owl, Grass	15.2	Owl, Typical	4.5
Owl, Hawk	15.2	Owl, Typical	15.2
Owl, Hissing	16.1	Owl, Typical/True	17.1
Owl, Hobby	16.1	Owl, Tyto	15.2
Owl, Hobgoblin	16.1	Owl, White	16.1
Owl, Hume's	17.8	Owl, White Breasted	16.1
Owl, Little, Rodrigues	4.5	Owl, Wood	15.2
Owl, Masked	15.2	Owl, Wood, Hume's	17.8
Owl, Mauritius	4.5	Owlets	15.2
Owl, Monkey-faced	16.1	Owlet, Forest	4.5
Owl, Night	16.1	Parakeet, Carolina	3.1
Owl, Pygmy	15.2	Parrots	3.3
Owl, Rat	16.1	Parrots	22.1
Owl, Saw-whet	15.2	Partridge	14.1
Owl, Scops	15.2	Partridge	14.2
Owl, Scops, Anjouan	4.5	Partridge, Arabian	14.2
Owl, Scops, Bruce's	17.4	Partridge, Barbary	14.2
Owl, Scops, Collared	15.1	Partridge, Grey	14.2
Owl, Scops, Common	17.5	Partridge, Philby's	14.2
Owl, Scops, Comoro	4.5	Partridge, Red-legged	14.2
Owl, Scops, Eurasian	17.3	Partridge, Rock	14.2
Owl, Scops, Mauritius	4.5	Partridge, Sand	14.2

Name	Par.	Name	Par.
Partridge, Sand	26.2.3	Pelican	26.3.1
Pheasants	14.1	Pelican, White	26.3.1
Peacock	14.5	Pheasant	1.3
Peacocks	14.5.1	Pheasants	13.1
Peafowl	14.5	Pheasants	14.5.1
Peafowl, Blue	14.5	Pigeon	18.1
Peafowl, Green	14.5.1	Pigeon	19.1
Peafowl, Indian	3.5	Pigeon, Blue-crowned	18.1
Peafowl, Indian	1.6.1	Pigeon, Crested	19.1
Peafowl, Indian	14.1, 5	Pigeon, Passenger	3.1
Peahen	14.5	Pigeon, Passenger	4.5
		Pigeon, Rock	19.3
		Pigeon, Wood, Common	19.3
		Partridge	13.1
		Partridge	1.3
		Partridge	14.1
		Partridge	14.2.1
		Partridge, Rock	14.2.1
		Pelican	4.2
		Pelican	26.3.1
		Pelican, white	26.3.1
		Pipits, Meadow	2.5
		Pittas	20.1
		Plover	26.3.2
		Plover, Egyptian	26.3.2
		Plover, Spur-winged	26.3.2
		Pygmy-Tyrant, Short-tailed	20.1
		Quail	13.1
		Quail	14.1
		Quail, Blue, Asian	14.1
		Quail, Common	14.3
		Quail, Common	14.3.1
		Quail, New World	13.1, 2

Palestine Sunbird (*Cinnyris oseus*)
(*Photo: Dr. Marcel Holyoak*) [4]

Name	Par.
Quelea, Red Billed	3.6
Raven	4.6.3
Rails	4.2
Rales	1.4
Rhea	7.1
Rockthrush, Blue	17.5.1
Rockthrush, Blue	23.2
Rook	22.2
Seabirds	17.2
Sandgrouse	14.2.1
Sandgrouse	18.1
Sandgrouse, Black-bellied	14.2.1

Name	Par.
Sandgrouse, Crowned	14.2.1
Sandgrouse, Lichtenstein's	14.2.1
Sandgrouse, Pin-tailed	14.2.1
Sandgrouse, Spotted	14.2.1
Secretary Bird	9.2
Shearwater, Manx	4.3
Skylarks	2.5
Solitaire	4.5
Solitaire, Reunion	4.5
Solitaire, Reunion	18.1
Solitaire, Rodrigues	4.5
Solitaire, Rodrigues	18.1
Sparrow, Dead Sea	23.2
Sparrow, House	23.1, 2
Sparrow, Rock	23.2
Sparrow, Spanish	23.2
Sparrow, Tree, Eurasian	23.1
Starling	2.7
Starling	3.6
Starling, European	4.3
Stork	4.2, 4
Stork	4.8
Stork	26.3.3
Stork, White	1.6.1
Stork, White	26.3.3
Strigids	17.1
Sunbirds	20.1
Swan	6.3
Swan	12.3
Swallow	1.3
Swallow	4.2
Swallow, Barn	21.1
Swallow, Chimney	21.1

Spanish Sparrow or Willow Sparrow
(Passer hispaniolensis)
(Photo: Dr. Marcel Holyoak) [5]

---

[5] This image is a reproduction of the photograph taken by Dr. Marcel Holyoak and used with permission. Originally placed on www.flickr.com, the original photograph can be viewed at: http://www.flickr.com/photos/maholyoak/5917420587/sizes/l/in/photostream/

Name	Par.	Name	Par.
Swallow, Common	21.2	Vulture	10.1
Swallow, European	21.1	Vulture	10.2
Swallow, Red-rumped	21.2	Vulture, American, Black	9.2
Swallow, Tree	21.1	Vulture, American, Black	10.2.1
Swift	4.2, 4	Vulture, Bearded	10.2.1
Swift	24.1	Vulture, Black	10.2.1
Swift, Alpine	25.1, 2	Vulture, Carrion	6.3
Swift, Common	11.1	Vulture, Egyptian	6.3
Swift, Little	25.2	Vulture, Egyptian	9.3
Swift, Palm, African	25.1	Vulture, Egyptian	10.2
Swift, Pallid	25.2	Vulture, Griffon	9.3
Swift, Typical	25.1	Vulture, Griffon	10.1, 2
Swift, White-collared	25.1	Vulture, Griffon, Himalayan	9.3
Swiftlet, Edible-nest	25.1	Vulture, King	10.2
Swiftlet, South-East Asian	4.5	Vulture, Lappet-faced	10.2
Tapacolo	3.2	Vulture, Lappet-faced	10.2.1
Teal	3.5	Vulture, New World	9.2
Tern	2.2	Vulture, Old World	9.2,3
Tern, Arctic	4.2, 3	Vulture, Palm-nut	10.2
Tern, Sooty	3.4	Vulture, Turkey	9.2
Thrush	17.3	Vulture, Turkey	10.2
Tit	17.8	Warbler, Aquatic	2.4
Tit, Blue	2.7	Warbler, Grasshopper	2.4
Tit, Pygmy	20.1	Warbler, Leaf	2.4
Treepies	22.1	Warbler, Willow	2.4
Tree swifts	24.1	Warbler, Willow	4.2
Turkey	4.5	Warbler, Wood	2.4
Turkey	13.1, 2	Warbler, Sedge	2.4
Turkey, Brush	13.1	Waxbills	20.1
Turkey, Ocellated	13.2	Weaver	3.6
Turkey, Wild	3.1	Weaver, Sociable	17.1
Turkey, Wild	13.2	Weavers, Social	17.1
Vulture	4.7	Wood Larks	2.5
Vulture	4.8	Woodpeckers	1.3
Vulture	9.1	Woodpeckers	17.1, 2

CHAPTER 38
UNIDENTIFIED BIRDS

38.1 Single Old Testament Occurrences.
(See par. 26.2.1)

TEXT	KING JAMES	LIVING BIBLE	JERUSA-LEM BIBLE	NEW INTERNA-TIONAL	AFRIKAANS 1933	AFRIKAANS 1983
1 Kng 4:23.	fatted fowl	plump fowl	cuckoos	choice fowl	vet ganse	vetpluimvee
Job 38:36	inward parts	inward parts	ibis	heart	wolke	ibis
Is 13:21	creatures	howlers	owls	jackals	uile	uile
Is 34:15	great owl	owl	viper	owl	pylslang	uile
Mi 1:16	eagle	-	vulture	vulture	aasvoël	aasvoël

CHAPTER 38
UNIDENTIFIED BIRDS

38.2 Double Old Testament occurrences.
(See par. 26.2.2)

TEXT	KING JAMES	LIVING BIBLE	JERUSA-LEM BIBLE	NEW INTERNA-TIONAL	AFRIKAANS 1933	AFRIKAANS 1983
Deut 14:13	vulture	-	-	falcon	gier	roofvoëlsoorte
1Sam 26:20	partridge	partridge	partridge	partridge	patrys	patrys
1Kng 10:22	peacocks	peacocks	baboons	baboons	poue	bobbejane
2Chr 9:21	peacocks	peacocks	baboons	baboons	poue	bobbejane
Ps 84:4	swallow	swallow	swallow	swallow	swaweltjie	swaeltjie
Prv 26:2	swallow	swallow	swallow	swallow	swaweltjie	voëltjies
Is 34:15	vultures	kites	kites	falcons	kuikendiewe	aasvoëls
Is 38:14	swallow	dove	dove	mourning dove	kraanvoel	duif
Is 38:14	crane	swallow	swallow	swift/thrush	swaeltjie	voël wat kerm
Jer 8:7	swallow	swallow	crane	thrush	kraanvoël	kraanvoël
Jer 8:7	crane	crane	swallow	swift	swaeltjie	swaeltjie
Jer 17:11	partridge	bird	partridge	partridge	voël	fisant

CHAPTER 39
"UNCLEAN" BIRDS

[13]And these are they which ye shall have in abomination among the fowls; they shall not be eaten, they are an abomination: the eagle, and the ossifrage, and the ospray, [14]And the vulture, and the kite after his kind; [15]Every raven after his kind; [16]And the owl, and the night hawk, and the cuckow, and the hawk after his kind, [17]And the little owl, and the cormorant, and the great owl, [18]And the swan, and the pelican, and the gier eagle, [19]And the stork, the heron after her kind, and the lapwing, and the bat.

Lev 11:13–19 (KJV)

39.1 Leviticus

KING JAMES VERSION	LIVING BIBLE	JERUSALEM BIBLE	NEW INTER-NATIONAL	AFRIKAANS 1933	AFRIKAANS 1983
eagle	eagle	tawny vulture	eagle	arend	arende
ossifrage	vulture	griffon	vulture	lammervanger	lammervangers
osprey	osprey	osprey	black vulture	aasvoël	visarende
vulture	falcon	osprey	red kite	kuikendief	aasvoël
kite	kite	kite	black kite	valk	valksoorte
raven	raven	buzzard	raven	kraai	kraaisoorte
owl	ostrich	raven	horned owl	volstruis	volstruise
nighthawk	nighthawk	ostrich	screech owl	naguil	nagvalke
cuckoo	sea-gull	screech owl	gull	seemeeu	seemeeue
hawk	hawk	sea-gull	hawk	kleinvalk	roofvoëls
little owl	owl	hawk	little owl	steenuil	uile
cormorant	cormorant	horned owl	cormorant	visvanger	visvangers
great owl	ibis	cormorant	great owl	steunuil	ibises
swan	marsh hen	barn owl	white owl	silweruil	flaminke
pelican	pelican	ibis	desert owl	pelikaan	pelikane
eagle	vulture	pelican	osprey	klein-aasvoël	swane
stork	stork	white vulture	stork	springkaanvoël	ooievaars
heron	heron	stork	heron	reier	reiers
lapwing	hoopoe	heron	hoopoe	hoep-hoep	hoep hoepe
bat	bat	hoopoe	bat	vlermuis	vlermuise
		bat			

CHAPTER 39
"UNCLEAN" BIRDS

> [12]But these are they of which ye shall not eat: the eagle, and the ossifrage, and the ospray, [13]And the glede, and the kite, and the vulture after his kind, [14]And every raven after his kind, [15]And the owl, and the night hawk, and the cuckow, and the hawk after his kind, [16]The little owl, and the great owl, and the swan, [17]And the pelican, and the gier eagle, and the cormorant, [18]And the stork, and the heron after her kind, and the lapwing, and the bat.
>
> **Deut 14:12-18 (KJV)**

39.2 Deuteronomy

KING JAMES VERSION	LIVING BIBLE	JERUSALEM BIBLE	NEW INTERNATIONAL	AFRIKAANS 1933	AFRIKAANS 1983
eagle	eagle	tawny vulture	eagle	arend	arende
ossifrage	vulture	griffon	vulture	lammervanger	lammervangers
osprey	osprey	osprey	black vulture	aasvoël	visarende
glede	buzzard	kite	red kite	kuikendief	swart aasvoël
kite	kite	buzzard	black kite	valk	valke
vulture	-	-	falcon	gier	roofvoëls
raven	raven	raven	raven	kraai	kraaisoorte
owl	ostrich	ostrich	horned owl	volstruis	volstruise
nighthawk	nighthawk	screech owl	screech owl	naguil	nagvalke
cuckow	sea-gull	sea-gull	gull	seemeeu	seemeeue
hawk	hawk	hawk	hawk	kleinvalk	roofvoël
little owl	screech owl	owl	little owl	steunuil	uile
great owl	great owl	barn owl	great owl	steunuil	ibises
swan	horned owl	ibis	white owl	silweruil	flaminke
pelican	pelican	pelican	desert owl	pelikaan	pelikane
eagle	vulture	white vulture	osprey	klein-aasvoël	swane
cormorant	cormorant	cormorant	cormorant	visvanger	visvangers
stork	stork	stork	stork	springkaanvoël	ooievaars
heron	heron	heron	heron	reier	reiers
lapwing	hoopoe	hoopoe	hoopoe	hoep-hoep	hoep hoep
bat	bat	bat	bat	vlermuis	vlermuise

CHAPTER 40
BIRDING IN ISRAEL

> "The king of Israel has come out to look for a flea
> as one hunts a partridge in the mountains."
>
> **1 Sam 26:20 (NIV)** [1]

40.1 Introduction

I have often been asked: "If there are still so many species to be seen and places to visit in your own country, why go birding in Israel?" I believe that the uniqueness of this tiny area lies in the fact that the Middle East is the region in the world where a large part of its history had been extensively documented and preserved in the Hebrew Canon and other sources. Coupled with the fact that the Hebrew Canon forms such a crucial part of two of the major religions of the world, it produces an unequalled birding experience to the believer as well as the non-believer.

A short overview on the present day birding situation in Israel is given for the benefit of future visitors. The objective of this section is in no way an attempt to serve as a visitor's guide. As I am not tied to or involved with anyone in Israel, so I

African Spotted Eagle Owl *(Bubo africanus)*.
(Photo: Rus Koorts) [2]

---

[2]   This image is a reproduction of the photograph taken by Rus Koorts from Pretoria, South Africa and used with permission. Originally placed on www.flickr.com, the original photograph can be viewed at: http://www.flickr.com/photos /ruslou/404468559/.

believe that I am able to give a more objective assessment of the situation, as those found in the tourist brochures. Having said that, I must say that I would urge anyone who has the means and/or opportunity to visit the area to do so.

The two books by Shirihai, et al, (Refer to par. 29.4), have extremely detailed information on each and every birding spot, and are highly recommended. I have selected about 30 of the best-known spots, which I have visited and found easily accessible driving a two wheel drive saloon car, and which cover the entire country. These spots could all be visited within about ten birding days, but as time is usually at a premium when visiting a foreign country, and one has to decide on a limited number of spots, I have opted for a three star

Egyptian Goose *(Alopochen aegyptiacus)*
(Photo: Ted van den Bergh) [3]

rating system of the sites, to serve as some guideline should one need to prioritize them. These are my personal choices, and I should stress that to me, any place in Israel is special. I would refrain from naming species to be found at specific spots, as I found the most interesting species in the most unlikely spots, and vice versa. The system is:

#:   An absolute must of which there are ten sites.
#:   Nice to visit.
#:   Could be passed.

The Sabbath is a crucial part of the social structure in Israel. It starts at 18h00 on Friday evenings, and lasts for twenty-four hours to 18h00 on Saturday evenings. Jewish businesses operate normally during the remaining times of the week. During the Sabbath however, all activities owned or managed by Jews or the State of Israel come to a complete standstill. This includes public transport, offices and businesses. Take note that during the Sabbath certain areas are closed to traffic. As these areas are not always clearly marked, one should take great care when traveling on the Sabbath. I have found for myself how this can turn into an extremely unpleasant (but in retrospect, comical) experience.

---

[3] This image is a reproduction of the photograph taken by Ted van den Bergh and used with permission. Originally placed on www.flickr.com, the original photograph can be viewed at: http://www.flickr.com/photos/webted/2521960644/sizes/l/in/set-72157605244874930/

40.2 Chutzpah

Following is an explanation of this term from the website www.dictionary.reference.com. *Chutzpah* is the quality of audacity, for good or for bad. The word derives from the Hebrew word *hutspa*, meaning "insolence", "audacity", and "impertinence." The modern English usage of the word has taken on a wider spectrum of meanings, having been popularized by vernacular use, film, literature, and television.

In Hebrew, *chutzpah* is used indignantly, to describe someone who has over-stepped the boundaries of accepted behavior with no shame. But in Yiddish and English, *chutzpah* has developed ambivalent and even positive connotations. *Chutzpah* can be used to express admiration for non-conformist but gutsy audacity. One common English adaptation of *chutzpah* is "hoodspa", which has a mostly positive connotation. Leo Rosten in *The Joys of Yiddish* defines *chutzpah* as "gall, brazen nerve, effrontery, and incredible 'guts'; presumption plus arrogance such as no other word and no other language can do justice to." In this sense, *chutzpah* expresses both strong disapproval and a grudging admiration. The word has also entered Polish from Yiddish and is written as *hucpa* in Polish. It likewise means arrogance, audacity and shamelessness.

I am sure that at least some Israeli Jews would admit that this would be an accurate description of their national mentality. It is a trait born out of necessity, as the modern State of Israel had been the scapegoat of virtually the whole international community, for as long as it has existed. I believe that many of the achievements in defending and developing the country, has been a direct result of this trait, as it enabled them on many occasions to achieve the seemingly impossible in battle and work. My experience is that this mentality manifests itself to some extent when an Israeli meets a stranger. But as soon as the bona fides of the stranger have been established, it changes into friendliness, warmth and hospitality. In other words, the visitor to Israel should not be deceived by first impressions. One might initially feel unwelcome, but an attempt to accept the people as they are, will eventually be richly rewarded.

Yellow Wagtail *(Motacilla flava)*
(Photo: Tony McLean) [4]

---

[4] This image is a reproduction of the photograph taken by Tony McLean and used with permission. Originally placed on www.flickr.com, the original photograph can be viewed at: http://www.flickr.com/photos/great_driffield/4526833300/.

40.3 Parks and Reserves

Israel has declared fourteen National Reserves (NR), which are more nature orientated, and thirty-eight National Parks (NP), which offers protection to nature but often, in addition, include a historical site e.g., Caesarea, Mount Masada, Qumran, Ein Avdat, etc. All these locations have proved to be productive birding spots, and should always be visited if possible. They are popular outdoor destinations for the local population, and should therefore preferably not be visited on public holidays.

Green Bee-Eater *(Merops orientalis)* *(Photo: Gary Knrd)* [5]

40.4 Infrastructure

Israel is a first world country, with the accompanying infrastructure, but extremely small. It is only in the far south were one would have to travel more than fifty kilometers between birding sites. Otherwise distances in the order of twenty kilometers or even less are common. One hardly ever needs to travel on gravel roads, and accommodation facilities are plentiful and cover the whole spectrum, from backpacking hostels to Bed and Breakfast guesthouses to luxury hotels. Generally, I have found the accommodation offered on kibbutzim acceptable and practical as they are usually good birding sites in themselves or close to such sites. The staff members are always obliged to make special arrangements to suit birders' special needs.

40.5 Military Zones

Since its proclamation on 14 May 1948, by the late David Ben Gurion, the security of the State of Israel has been under constant threat. For this reason the Israeli Armed forces have been playing a larger part in the everyday life in the country as in most other. When birding in Israel one soon becomes aware of this fact as large areas are controlled by the

---

[5] This image is a reproduction of the photograph taken by Gary Knrd. and used with permission. Originally placed on www.flickr.com, the original photograph can be seen at: http://www.flickr.com/photos/avianphotos/5348568027/.

Rock Dove / Rock Pigeon *(Columba livia)*
(Photo: Shlomi Chetrit) [6]

military, often restricting access to areas where one would like to bird. I have however found that should one refrain from fanatical twitching, this poses no problem. In such cases it is wise to adhere to the age-old rule: "if in doubt, don't." Any good map of the country, even those supplied by car rental companies, are adequate. The variety of birds to be seen in accessible areas is more than gratifying.

As is normally the case when traveling abroad, but more so in the Middle East, ensure that each and every individual in the party has his/her passport on his/her person at all times. Keep in mind that people moving about with binoculars, telescopes and telescopic camera lenses are prone to draw the attention of the military and/or other security personnel. It is always useful to keep those field guides handy to verify or strengthen your explanations. It should be needless to say that the above is paramount at border crossings and/or where customs officials are operating. One should at all times be prepared to be confronted by security personnel at road blocks, on a side walk, while strolling in the park, or in the remotest part of the desert. Keep in mind that the person, who you may be confronted by, may have absolutely no idea what birding is about or who or what a birder is.

40.6 Northern Region

Ma'agan Mikhael (# # #). The area has many fish ponds dotted along the coast.

The Jezreel Valley (#). The areas between agricultural fields are capable of being quite productive.

---

[6]   This image is a reproduction of the photograph taken by Shlomi Chetrit and used with permission. Originally placed on www.flickr.com, the original photograph can be viewed at: http://www.flickr.com/photos/shlomz/5818146662/in/set-72157626804218111.

The Hula Valley (# # #). The area surrounding the Hula NR proved much more productive to me, as it contains secluded marshy areas which one should in any case always be on the lookout for when traveling.

Hula NR (#). The Hula NR proper was disappointing, as the reeds are extremely dense, the water deep, and the tourists many.

Banias (#). The climate is turning cold.

Dan (#). The forests are unique.

Mount Hermon (#). Better known as a ski resort.

Black-necked Stilt *(Himantopus mexicanus)*
(Photo: David Hofmann) [7]

The Golan Heights (# #). The high altitude pastures form unique habitats.

The Sea of Galilee (# # #). The coastline is easily reached at several points.

Gamla NR (#). The only breeding site of Griffon Vultures *(Gyps fulvus)* in Israel.

Tiberias (# # #). The harbor is always alive with gull and waterfowl.

Biet She'an (#). The town proper is a must from a historic point of view, but the prime birding spots are at the fish ponds in the surrounding areas.

Kibbutz KfarRuppin (# # #). The fish ponds on the kibbutz are one of the most productive birding sites in Israel. The accommodation and dining facilities are superb, and the birding spots and ringing station are within walking distance from the village.

---

[7]   This image is a reproduction of the photograph taken by David Hofmann and used with permission. Originally placed on www.flickr.com, the original photograph can be viewed at:http://www.flickr.com/photos/23326361@N04/4248891705/.

Kibbutz Hamadya (# #). The fish ponds are situated next to route 90.

Gan Ha Shalosha (#) In the Beit She'an district.

Mount Gilboa (#). Marvelous sight seeing drive.

40.7 Central Region

Jerusalem (#). The city has many parks, boulevards and tree rich areas that are ideal for birding. The following are a few examples:

Independence Park (#). The park is situated in the new city center.

Mount Herzl (#). On a major bus route, adjacent to Holocaust Museum.

Biblical Zoo (#). Situated on the outskirts of the city.

Wolfson Garden (#). Adjacent to the Old City Walls.

Tel Aviv (#). The beach-front area is always alive with bird life.

Jaffa (#). The ancient city of Jaffa is a popular tourist site on the coast.

Yardon River (#). Flowing into the Mediterranean.

The Dead Sea (# # #). One of the sparrows of the Middle East has been named in the vernacular after this natural wonder, and its generic name refers to the area to the East of the sea. The Dead Sea Sparrow *(Passer Moabiticus)* is a common resident in the area.

Ein Gedi (###). Mentioned in the Bibles as one of the hide-outs of the young man David when he fled from King Saul.

Ein Bokek (#). The renowned tourist center consists of a number of luxury hotels and beach resort.

40.8 Southern Region

Eilat (# # #). The most southern center in Israel is situated on the northern tip of the Gulf of Aqaba. At least nine spots, including the bird sanctuary and ringing station, are within easy reach from the city center.

Km 20 and 33 (# #). There are some excellent birding spots adjacent or close to the main road.

Jotvata (# #). Usually a refreshment stop for travelers to Eilat, the village has some excellent birding spots close to the main road, behind the gas station and in the kibbutz fields,

Black-crowned Night Heron *(Nycticorax nycticorax)*
(Photo: Dr. Marcel Holyoak) [8]

Kibbutz Lotan (# # #). Situated on the main route southwards to Eilat, this Kibbutz has given birding tourism a high priority in their activities.

Shizafon Sewerage Works (# #). On route 40 towards Mitzpeh Ramon.

Hal Bar NP (# #). This National Park is a tourist attraction but includes some excellent desert birding.

Timna NP (# #). This National Park is a tourist attraction but includes some excellent desert birding.

Mitzpe Ramon (#). The desert has many surprises.

---

[8] This image is a reproduction of the photograph taken by Dr. Marcel Holyoak. Originally placed on www.flickr.com, the original photograph can be viewed at: http://www.flickr.com/photos/maholyoak/5917466691/sizes/l/in/photostream/.

Tsede Boker(#). The gardens at Ben Gurion's grave site and desert home.

Ein Avdat NP (# # #). This National Park is teeming with vultures and other raptors.

Nizzana (#). A long hot drive from Eilat or Kibbutz Lotan to the Israel- Egypt border.

Malachite Kingfisher *(Alcedo cristata)*
(Photo: Rus Koorts) [9]

---

[9]   This image is a reproduction of the photograph taken by Rus Koorts from Pretoria, in South Africa and used with permission. Originally placed on www.flickr.com, the original photograph can be viewed at: http://www.flickr.com/photos/ruslou/5001228090/in/set-72157624830170739

PSALM 8

**For the director of music. According to gittith.
A Psalm of David.**

LORD, our Lord,
how majestic is your name in all the earth!

You have set your glory in the heavens.
Through the praise of children and infants
you have established a stronghold against your enemies,
to silence the foe and the avenger.
When I consider your heavens,
the work of your fingers,
the moon and the stars,
which you have set in place,
what is mankind that you are mindful of them,
human beings that you care for them?

You have made them a little lower than the angels,
and crowned them with glory and honor.
You made them rulers over the works of your hands;
you put everything under their feet:
all flocks and herds, and the animals of the wild,
the birds in the sky, and the fish in the sea,
all that swim the paths of the seas.

LORD, our Lord,
how majestic is your name in all the earth!

Christiaan (Tian) Hattingh was born and bred in South Africa. In 1974 he completed a four year training course as a Forest Manager from the Saasveld Forestry College, now part of the Nelson Mandela Metropolitan University. In 1987 he completed a B.A. degree in Psychology and Philosophy at the University of South Africa. He studied Biblical Hebrew for three years at the same university, and later conducted Technicon classes to beginners. He has been an avid birder for the past 37 years, and is a founding member of the Rustenburg branch of BirdLife South Africa. He has been birding in Botswana, Zimbabwe, Namibia, Malawi, China, and Thailand. He has visited Palestine several times, studying the birds of that region, and has presented talks on "Birding in Israel". In 2002 he moved to mainland China, where he is an ESL teacher to this day. Personal website: www.tianhattingh.com